Knowing Her Place

NEW HORIZONS IN MANAGEMENT

Series Editor: Professor Sir Cary L. Cooper, CBE, *50th Anniversary Professor of Organizational Psychology and Health at Alliance Manchester Business School, University of Manchester, UK and President of the Chartered Institute of Personnel and Development and British Academy of Management*

This important series makes a significant contribution to the development of management thought. This field has expanded dramatically in recent years and the series provides an invaluable forum for the publication of high quality work in management science, human resource management, organizational behaviour, marketing, management information systems, operations management, business ethics, strategic management and international management.

The main emphasis of the series is on the development and application of new original ideas. International in its approach, it will include some of the best theoretical and empirical work from both well-established researchers and the new generation of scholars.

Titles in the series include:

Knowing Her Place

Positioning Women in Science

Valerie Bevan

Honorary Teaching Fellow, Lancaster University Management School, UK

Caroline Gatrell

Professor of Organization Studies, University of Liverpool Management School, UK

NEW HORIZONS IN MANAGEMENT

Edward Elgar
PUBLISHING

Cheltenham, UK • Northampton, MA, USA

Published by
Edward Elgar Publishing Limited
The Lypiatts
15 Lansdown Road
Cheltenham
Glos GL50 2JA
UK

Edward Elgar Publishing, Inc.
William Pratt House
9 Dewey Court
Northampton
Massachusetts 01060
USA

Paperback edition 2019

A catalogue record for this book
is available from the British Library

Library of Congress Control Number: 2017947078

This book is available electronically in the **Elgar**online
Business subject collection
DOI 10.4337/9781783476527

ISBN 978 1 78347 651 0 (cased)
ISBN 978 1 78347 652 7 (eBook)
ISBN 978 1 78990 426 0 (paperback)

Typeset by Servis Filmsetting Ltd, Stockport, Cheshire
Printed by CPI Group (UK) Ltd, Croydon CR0 4YY

For women scientists, past, present and future

For Byron and Henry
Valerie Bevan

For Tony, Anna and Emma
Caroline Gatrell

Contents

About the authors

Valerie Bevan is an Honorary Teaching Fellow at Lancaster University Management School, UK. She is a microbiologist who has worked in public sector organizations including the NHS where her main national and international contribution has been leading the development of standardized methods in diagnostic microbiology.

For many years during Valerie's long career she was steeped in the expectation that she should 'know her place' and not threaten the status quo. She came to feminist and critical management studies late when undertaking a management course at the University of York, followed by a PhD in Management Learning and Leadership at Lancaster University. It was during this time that she came to realize that science itself could be questioned and she found a new freedom to challenge how accepted masculine scientific norms influenced the multiethnic workforce where women were in the numerical majority but few made it to the top jobs.

Valerie has been a keynote speaker on the subject of women in science many times and has facilitated at various workshops. She is also a member of the Advisory Board to the Critical Studies Research Group at Durham University. Valerie has been a member of the council of the Institute of Biomedical Science, and has contributed to the Science Council on diversity and equality. She currently chairs the British Society for Microbial Technology.

Caroline Gatrell is Professor of Organization Studies at University of Liverpool Management School, UK. Caroline's research centres on family, work and health. From a socio-cultural perspective, Caroline examines how working parents manage boundaries between paid work and their everyday lives. In so doing she explores interconnections between gender, bodies and employment, including theorizing on masculinity and employment, as well as development of the concepts of 'maternal body work' and 'pregnant presenteeism'.

Her work is published in leading journals including: *Human Relations*; *British Journal of Management*; *Gender, Work & Organization*; *Social Science & Medicine*; *International Journal of Management Reviews*; and *International Journal of Human Resource Management*.

Caroline enjoys teaching research development and capacity building among PhD students and early-career scholars.

Figures

Tables

Preface

This book has been almost ten years in the making. Following initial collection and analysis, we took a step back from our data. We sought to attain critical distance, enabling us to see the bigger picture regarding women's position in science. We have spent time reflecting on how we might interpret our findings, and how we might best articulate and explain the persistent 'placing' of women scientists in roles that offer limited opportunities for research leadership.

As a team we bring together Valerie's insider knowledge as a senior scientist and Caroline's experience as a management sociologist. We believe this combination has allowed us to develop a framework and a perspective (see Figure 1.1 in Chapter 1) that have potential to shed light on women's status beyond the arena of science.

We suggest that Figure 1.1 'Knowing her place – positioning women in science' offers potentially wider application and anticipate that it may, in future, make a contribution more broadly in the context of research on women's work.

Valerie Bevan and Caroline Gatrell

Acknowledgements

There are many friends and colleagues who have assisted in the shaping of this book. In particular we would like to thank Byron and Henry Bevan, Tony Gatrell and Mark Learmonth for their critical reviews and insightful comments on earlier versions of the book as we developed our line of thought. We would also like to thank Maggie Vearey for reading the whole manuscript twice over for polish and proofing. We are grateful to Sarah Patterson for all her support over many years.

We owe a special debt to the anonymous scientists who kindly gave up their time to be interviewed as part of our research and without whom the book would not have been written.

Abbreviations

AAUW	American Association of University Women
AHSSBL	arts, humanities, social sciences, business and law
A Level	General Certificate of Education Advanced Level
AWIS	Association for Women in Science
BSc	Bachelor of Science
CEO	chief executive officer
DfE	Department for Education
EC	European Commission
ECU	Equality Challenge Unit
EHRC	Equality and Human Rights Commission
EU	European Union
F/T	full-time
GCSE	General Certificate of Secondary Education
HCPC	Health and Care Professions Council
HCS	healthcare scientist/science
HoCSTC	House of Commons Science and Technology Committee
IBMS	Institute of Biomedical Science
MBA	Master in Business Administration
MSc	Master of Science
NHS	National Health Service
OECD	Organisation for Economic Co-operation and Development
O Level	General Certificate of Education Ordinary Level, now GCSE

ONS	Office for National Statistics
PhD	Doctor of Philosophy
P/T	part-time
RC Path	Royal College of Pathologists
R&D	research and development
SET	science, engineering and technology
STEM	science, technology, engineering and mathematics
STEMM	science, technology, engineering, medicine and mathematics
STFC WiSTEM	Science and Technology Facilities Council, Women in Science, Technology, Engineering and Maths Network
UKRC	UK Resource Centre for Women
WISE	Women in Science and Engineering
WWC	Women & Work Commission

1. Introduction: Setting the scene

> [W]omen are operating within contradictory sets of meanings: 'contribute fully as equal members of the team' but 'remember your real place'.
>
> (Newman, 1995, p. 19)

When Valerie Bevan began her career as a junior biomedical scientist, she was discomfited to be expected to polish the desk of a senior male colleague. Under such circumstances, perhaps it is not surprising that she had to fight hard to advance her position as a serious scientist. Having achieved, at the end of her career, a director-level role, Valerie hoped that things might have improved for the younger women scientists following in her footsteps. Similarly, Caroline Gatrell, whose academic research on women in organizations tends to focus on women in their mid-thirties and forties, also wondered whether or not the situation for women in science might be changing for the better. Both of us were disappointed. As we show in this book, change has been slow and women's position in science remains tenuous.

We ask the questions: Why should men dominate the senior roles in the science workplace? What factors get in the way of women achieving as highly as men in science? And why should it be that, while there are some examples of women in senior management roles, few women are positioned at the leading edge of science – especially in research positions involving notions of 'creative genius' (Battersby, 1989, pp. 2–3)?

To attempt to answer these questions, we explore here how women experience scientific careers in healthcare science-related posts, the area most familiar to Valerie. We see healthcare science as a subset of the more general term 'science'. We do not profess that the conclusions we draw are applicable in all areas of science but we are confident from our own understanding and research that there are significant experiences encountered by our interviewees that are common to women who work in other branches of science. In developing our arguments, we offer new interpretations of the barriers and influences that confine women within operational and lower-level management positions, and deter or prevent women from reaching the most senior research roles.

Today, girls continue to outperform boys at General Certificate of Secondary Education (GCSE) level (Department for Education [DfE], 2015; Stubbs 2016) and at General Certificate of Education Advanced Level

(A Level) (DfE, 2015). More women than men study science at university, where statistics show that female science students consistently outperform males (Equality Challenge Unit [ECU], 2016; Ratcliffe, 2013). Despite this promising beginning, the attrition rate of postgraduate women progressing to senior roles in science is notable, and significantly fewer women than men hold prestigious roles in science (Greenfield, 2002a; Kirkup et al., 2010). For instance, at the highest level, out of a total of 1512 current Fellows (Honorary Fellows not Royal Fellows) elected to the Royal Society, only 144 are women (9.5 per cent) (Royal Society, 2017) and even in 2017, only 15 women of a total of 61 were elected (24.6 per cent) (Royal Society, 2017). Furthermore, of the science-related Nobel Prizes, one woman won a prize in 2015 and one in 2014 (each in 'Physiology or Medicine'), compared with seven and eight men respectively. No awards were made to women in 2016.

In this book, we identify the manner in which women are allocated a 'place' within healthcare science professional and structural hierarchies. Exploring the notion of 'place' as a gendered concept, we note that women in science continue to be excluded from the most prestigious 'place[s]' of scientific 'action', which tend to be populated by men (Miller, 1986, p. 75). In contrast, women's 'place' is usually located at a distance from leading positions within science. We show how male research leaders in science, both consciously and unconsciously, draw on notions of 'difference' and 'otherness' to position women in lower grades. For the majority of women scientists their place is frequently in the background, supporting the endeavours of (often male) creative research scientists. Women's place in science is still defined by expectations that they should support male colleagues, often at the expense of their own careers.

The hierarchical professional structure in healthcare science is congruent with the demarcation strategies described by Witz (1992) in relation to radiography. At the top are powerful medically qualified doctors who usually lead such laboratories in the public sector. Then come clinical scientists or research healthcare scientists who are placed above biomedical scientists, the third group in the hierarchy (Valerie's professional background). Research healthcare scientists and biomedical scientists often have similar roles and responsibilities and the pay may be similar but the training and career opportunities are different.[1]

Within this structure, more women than men occupy the lower operational roles and men tend to rise to the more senior roles (Bevan, 2009; Greenfield, 2002a; Kirkup et al., 2010). We suggest that organizational assumptions regarding women's lowered place relative to men in science, remain stable and robust. This is despite the presence, over more than 20 years, of focused and informed initiatives seeking to disrupt the status quo (see, for example, Greenfield, 2015).

We argue here that the 'placing' of women in subordinate roles leads to a conflict between the desire among some women to further their own scientific interests, and subtle yet powerful professional beliefs about where, and how, female scientists should be treated and positioned within the hierarchical professional structure. We make these observations drawing upon data from qualitative interviews undertaken between 2005 and 2015 among 47 women and men working in healthcare science-related posts in the UK, of whom 42 worked (or had worked) in research and development (R&D) posts for a major part of their role (see the section on research design below and the Appendix for more details on their backgrounds and roles). The views and experiences of the interviewees illuminate how healthcare scientists, including women themselves, appear often to share (or become resigned to) common perceptions about the location of women's 'real place' in science compared with men (Newman, 1995, p. 19). These long-established perceptions are consolidated by structures and behaviours that continuously reproduce and perpetuate situations whereby women's place continues to be located firmly within the lower ranks in science.

Our research affirms that the most prestigious leading and creative research roles within healthcare science in the UK are habitually 'reserved' for male scientists (often those who are white and from privileged backgrounds, as in politics; see Puwar, 2004). By contrast, few 'VIP' roles in science are offered to female scientists. It remains unusual for women (or for workers with ethnic minority backgrounds) to be placed in leading and creative R&D posts. Rather, it tends to be the case that women populate the places reserved in the 'second class' operational arenas in science, usually in those supporting or administrative roles that offer male 'creative' research scientists the space to spread their wings.

As a consequence of tendencies within science to define women's 'place' as accommodating (rather than leading), we observe how aspiring women healthcare scientists – from their teenage years onwards – experience opportunities differently from their male counterparts. Our investigations show how girls are often offered inappropriate advice at school, perhaps due to unspoken assumptions that they don't need to know things that might take them to the top of their profession as they will never get that far. Young women may, as a consequence, be directed into operational, service and pastoral roles on entering science (where equivalent males might be offered opportunities and finance to develop research careers). Where women demonstrate ability and ambition, they may be persuaded into administrative careers, leaving the world of research for senior management roles (albeit in a science context). Further, women are often sidelined early in their careers due to their potential for maternity. Their capacity for reproduction (even if they never become mothers) is sufficient to invoke often unspoken, yet powerful, organizational assumptions that women's work orientations are necessarily lowered as a result of the possibility that they might bear children (Gatrell, 2013).

The strength and longevity of established hierarchies within the professional structures in science makes it hard for women to fight their way out of the lower-level places they have been allocated. Our research demonstrates how, even if women do escape from their designated 'place' (especially if they step into influential arenas where the decisions are made), such attempts may invoke opprobrium from colleagues (and sometimes their own families). As a result, women who appear to infiltrate the creative research arenas in science more usually reserved for men may be treated as 'space invaders' (Puwar, 2004, p. 10), usurpers or outsiders who find some doors remain firmly closed to them. For example, when the internationally successful neuroscientist Susan Greenfield was nominated to join the long list to become a Fellow of the Royal Society, she was refused admission to the prestigious halls of membership (Blackstock, 2004).

In what follows, we put forward a range of contributory reasons for why women's 'reserved' place in science is located so firmly in the second tier. Drawing upon the experiences and views expressed by our interviewees, we explore why women scientists continue to encounter fewer opportunities than men to become leading researchers. In doing so we identify four mechanisms that we believe establish and maintain women's lowered position in science compared with men and that we illustrate in the framework in Figure 1.1. These mechanisms combine to perpetuate and strengthen the subtle yet powerful social structures that consolidate the means by which women come to know (and are kept within) their 'real place' in science (Newman, 1995, p. 19).

We define the four mechanisms as:

- *subtle masculinities*, whereby masculine cultures in science tend to privilege men and marginalize women;
- *secret careers*, whereby subtle masculinities may invade heterosexual households in which women hide their career aspirations from their husbands/partners (all women in heterosexual partnerships in our study were, or had been, married; one woman was in same-sex civil partnership);
- *the concept of creative genius*, which is associated with male bodies, meaning that women find it hard to envision themselves (or be envisioned by others) in such a role;
- *m[o]therhood*, in which women's potential for maternity positions them as different and 'other'.

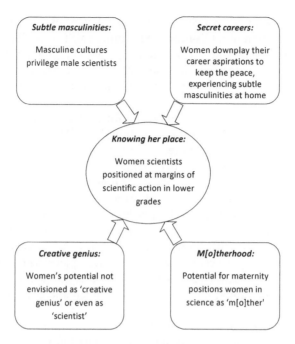

Figure 1.1 *Bevan and Gatrell Framework: Knowing her place – positioning women in science*

Below, we offer a brief description of each mechanism. We then, in Chapter 2, outline as background for our discussions the literature and narratives that contextualize the disadvantageous position of women in science. In doing so, we reflect upon the experiences of Rosalind Franklin and Susan Greenfield, both scientists who have been undervalued by their contemporaries in the world of science.

SUBTLE MASCULINITIES

As an explanation for how 'subtle masculinities' exclude women scientists from the most prestigious roles, we suggest that the world of science has long favoured 'hegemonic' forms of masculinity (Collinson and Hearn, 1996; Connell, 2005). By 'hegemony' we mean the ascribing of status to men within a male-dominated profession, where organizational and cultural practices privilege certain forms of masculinity and women are positioned in supporting roles. As we demonstrate in Chapter 6 with Alec's story, not all men find it easy to comply with the conventional image of hegemonic masculinity, which tends to imply heterosexuality, good health and the assumption of a full-time breadwinner role (Coles, 2008; Connell, 2005). For those men who are able to meet these 'hegemonic' criteria, the opportunities for career progression within science are far greater than for women (or for male scientists who fall outside notions of a masculine ideal).

We interpret hegemony in science as a form of subtle masculine influence. We show how such 'subtle masculinities' (see Bevan and Learmonth, 2013) act in the workplace to preserve senior positions within science for men, while women are expected to remain in their 'place' in the lower hierarchies (Newman, 1995). We illustrate how hegemonic cultures within science are taken for granted by both women and men. For this reason, the effects of powerful yet largely unacknowledged 'subtle masculinities' on women's careers in science may go unchallenged. Our research findings show how women in science often seek to avert the potential conflict and disturbance that might occur if they refuse their 'place' as operational or supporting, rather than leading scientists. As a consequence, we observe how very able and ambitious women may dampen and conceal their own aspirations.

A woman's role as a creative scientist is thus constrained so that women are steered away from their science-based focus towards administrative and lower-ranking managerial positions, often on the edges of science – for example, managing 'quality'. In what follows, we render visible the complex interactions between women and men within hegemonic workplace settings where men continue to have the advantages. Throughout the book we confront the failure to even out the unequal opportunities in healthcare science that are highlighted in mainstream literature on scientific life since the 1980s, urging that 'serious' notice be taken:

> [. . .] not only of the fact that science has been produced by a sub-set of the human race – that is, almost entirely by white, middle-class men – but also of the fact that it has evolved under the formative influence of a particular ideal of masculinity [associated with] 'virile' power. (Keller, 1985, p. 7)

SECRET CAREERS: SUBTLE MASCULINITIES AT HOME

We show that women are subject not only to the subtle masculinities at work but these may extend into the home spheres as well. Among women who are highly committed to their careers, some are constrained by a perceived need to remain invisible at home as a career woman in an effort to avoid conflict not only with male bosses at work but, for those in heterosexual relationships, sometimes even their husbands/partners. While some of our respondents seemed to have developed such strategies unconsciously, others reported proactively balancing a strong career focus with the requirement to appear less work oriented than they actually felt.

Among some heterosexual women with a strong commitment to their paid work and career, a need was expressed to recognize the ultimate importance of the career of the husband above her own. Such women not only took on the bulk of the home and child management, protecting their husbands from worrying about family responsibilities, but at the same time, they also released men from the need to be concerned about the career aspirations of their wives. Women did not ask their husbands to take a full share of the 'second shift' of domestic and care arrangements (Hochschild, 2003) but seemed resigned to the 'help' that they got from their spouses as being as much as they could expect. We suggest that women 'talked up' the importance and income of the spouse to rationalize why they took on more of the second shift than their husbands. However, even when women earned more than their husbands, some still protected the narrative of husbands/partners as main breadwinners, perhaps to sustain masculine identities.

We argue then, that women healthcare scientists may endure their subordinate place in the home, just as they do at work, to avoid conflict (Miller, 1986). Men have no need to change the status quo that gives them an advantage and in our study they were not challenged to do so by their wives.

CREATIVE GENIUS

The most prestigious positions or places within science globally are held by individuals to whom the description 'creative genius' may be applied and this title is invariably applied to men. The concept 'genius', as feminist philosopher Battersby (1989) points out, is not a factual term but is 'descriptive', involving judgement on the part of the assessor (p. 124). Genius is closely associated with science and the image of the creative or scientific genius is predominantly male and associated with names such as Aristotle, Newton and Einstein (Simonton, 1988). As Battersby (1989) has observed, the position of brilliant research scientist has historically been held by men, with women scientists either written out of science histories or positioned as helpers whose 'place' is located within a supportive trajectory – tidying the laboratory, taking and filing notes and undertaking experiments only under the supervision of great men.

Historically, women were not regarded as being capable of being a genius and for a woman to be seen as such was against the laws of nature or even 'loathsome' and 'aping the genius of the male – who is her (and nature's) lord and master' (Battersby, 1989, p. 77, citing Kant, 1790 [1951]). Although many of the qualities attributed to a genius (such as 'emotion' and 'imagination') are often regarded as feminine characteristics, women were, and remain, excluded from the accolade of genius (Battersby, 1989, p. 3).

Traditionally and to date, women scientists are embodied as 'other' and situated in a 'place' that effectively ensures that they may be occluded from social definitions of creative genius. Becoming a 'scientific genius', we suggest, may be as much about being given opportunities to thrive and shine as it is about innate brilliance. We argue that this historical legacy continues to influence how women are judged, meaning that in science professions, recognition of women's potential to become a scientific genius is commonly disregarded and their opportunities are limited compared with those on offer to men at equivalent levels.

M[O]THERHOOD

Genius is not the only arena in which the 'othering' of women has a negative influence on where they are 'placed' in science. It has long been acknowledged within gender studies that motherhood has a detrimental influence on women's careers (Blau et al., 2014; Gatrell 2008). We suggest, however, that childcare responsibilities are not in themselves a cause of women's subordinate position (although women in heterosexual relationships are more often than not responsible for the greater burden of childcare and housework). Rather, the predominant hegemonic masculinities within science serve to marginalize not only mothers, but also those women with potential to become mothers, due to unfair and unsubstantiated assumptions that childbearing reduces women's commitment to their career and dulls their ability to undertake creative or groundbreaking research over a lifetime's work.

In other words, women become identified as 'other' or more appropriately 'm[o]ther' (see Fotaki, 2011 and Cooper, 1992 for a discussion on the othering of mothers) and are subtly marginalized from opportunities to pursue research careers in science. As we explore the experiences of our research participants, we observe how motherhood is closely interlinked with women's exclusion from the highly esteemed roles in science. We further note how family commitments may even be detrimental to men in situations where caring responsibilities are visible.

RESEARCH DESIGN

We chose in-depth, semi-structured interviews based around life histories of 47 interviews among 33 women and 14 men working in healthcare-related science from both public (41) and private (6) sectors. A summary of the details about the interviewees may be found in the Appendix. In summary, all had worked in laboratories for part or all of their careers but 34 had moved into director, management or other posts whilst remaining practising scientists. Most (42) had worked for part or the whole of their career in R&D posts. Many had worked in microbiology and public health laboratories, Valerie's speciality, but for the purposes of this book, for the sake of preserving participants' anonymity, the more general term 'healthcare science' laboratories is used for all. The interviewees came mainly from two professional groups 'biomedical scientist' and 'clinical scientist' (which are statutory registered titles with the Health and Care Professions Council (HCPC)) [1]; and the more generic 'healthcare scientist'; four medically qualified doctors were also interviewed.

Some interviewees were dedicated research scientists, some were in laboratory manager or laboratory director positions, of whom most had international reputations. The majority of interviewees (41) were employed within different public sector organizations. Although the variety of the backgrounds of the interviewees meant that their career trajectories were different from each other, all shared some similar experiences within healthcare science.

Although the greater majority of the interviewees were healthcare scientists, the four medically qualified doctors were interviewed because of their key positions of influence – for instance, as deputy or chief executives or board members in the institutions where they and other interviewees worked. Medically qualified doctors are not the focus of this book but they play a significant role in public sector healthcare science/pathology laboratories, undertaking clinical responsibilities and developing and directing strategy. In most laboratories, they are at the top of the hierarchy, and control (usually from a distance) the work and careers of healthcare scientists, as became apparent in the interviews. It is in the context of the healthcare scientists that their comments are included, not in relation to their own careers.

Men as well as women were interviewed to enrich the data, as the views of men on the roles of women in science are important in the male-dominated science environment (Gatrell, 2006b; Pringle, 1998; Stanley and Wise, 1983). Interviewing men highlighted the perspectives that the men tended to have on their own and women's careers in contrast to the perspectives women have on their own careers. Masculine voices at the head of organizations, the way they viewed women's career progression, and the contrasts between the possibility and actuality of women rising to the top of organizations were highly relevant, taking note of the comment from Pringle (1998): 'I did not wish to attribute to the women perceptions or problems that are shared by both sexes' (p. 18).

Initial interviewees were Valerie's contacts in healthcare science and all other interviewees were found through 'snowballing', a process whereby interviewees suggest other possible interviewees (Whyte, 1993). All interviewees worked in the UK at the time of the interviews.

In-depth interviews were a suitable medium for communication and a way of encouraging people to make meanings of their experiences through a reflective process (Johnson, 2001). In-depth interviews with minimal structure using life histories (Atkinson, 2001) were our framework, allowing themes to emerge from the open-ended interviews and interactions between interviewer and interviewee (Holstein and Gubrium, 1997). Interviewees were invited to say what interested them about their work in science, to discuss the choices made and opportunities presented and taken, barriers faced and overcome (or not), their approach to their 'careers', and how they balanced work and home life.

All interviews were conducted by Valerie, who was often approached by colleagues within the healthcare science profession, knowing that this research was being undertaken, saying 'Can I come and see you?' or 'X would be such a good subject for you'. Each interview relationship was influenced by many factors including her female gender and history of many years working in a healthcare science environment. The interviewees were apparently frank and nearly all were reflective. Several interviewees changed their manner during the interview, starting by being hesitant but then, as the interview progressed, relaxing and speaking more readily. Several women followed this pattern, with an initial lack of acknowledgement that their inequitable experiences were the result of anything other than 'normal' behaviour by bosses, followed by the realization after reflection during the interview that these incidents may have been indicative of discrimination.

The interviews took place in a variety of locations, usually in offices, occasionally in hotels, over a ten-year period between 2005 and 2015, with some interviewees being interviewed more informally a second or third time to keep up to date with their progress. All interviews were recorded and subsequently transcribed by professional transcribers and each interviewee either chose or was assigned an 'anglicized' pseudonym to protect anonymity.

The interviewees spoke for as long as they wanted to. The contrast between interview transcripts is interesting in itself. Most interviews were between 60 and 90 minutes with the shortest two being between 40 and 45 minutes, and the two longest being over two-and-a-half hours. Each of the short interviews generated around 2800 words from the interviewee and 1000 from Valerie in the transcript, perhaps indicating a struggle to get the two women to talk. The longest interviews generated about 25 000 words and 800 from Valerie. These two women talked for long periods without interruption and Valerie oscillated between being polite and letting them continue and interrupting their train of thought to move them on to more pertinent topics. Each of these four interviews included a similar number of exchanges (around 90), indicating that Valerie's input was approximately the same irrespective of the contribution from the interviewees.

For three women, the interview was apparently near cathartic, as Haynes (2006) describes, and they shed tears as they recalled particular events. One woman was openly resentful of being unable to control her tears as she considered a man would do, because being tearful in her workplace might disadvantage her in her career. One man said 'But it's quite difficult . . . don't want to say . . . yeah I don't really want to go there'. After the interview he explained his feelings of discomfort away from the tape recorder, being assured that his reminiscences relating to difficulties in his workplace wouldn't be included in the research data.

The interviews were interpreted for the purposes of this book by both authors. Valerie brought to the analysis many years of working in a health-care science environment and witnessing first-hand the 'taken-for-granted' practices that had only recently become apparent to her as a new social and feminist researcher. She thus conducted the interviews from within the group being studied who worked mainly in public sector healthcare sciences. Caroline, a university professor experienced in qualitative research, specializing in the study of gender and work, brought to the analysis a history of working in both health management and academic settings. A close partnership developed between the authors over the ten years of working together on this research.

ETHICS

The heterogeneity of the interviewees and their variety of ethnic and social backgrounds was recognized both within and outside the study group. Thirteen of the interviewees (11 women, two men) had parents who were an ethnic minority in the UK, from seven countries across five continents: three women came to the UK specifically to study for a Doctor of Philosophy (PhD) degree. One man who was born in the UK undertook his secondary education and first degree outside Europe before returning.

Our research was undertaken outside the UK's National Health Service (NHS) and before interviewing began agreement was gained from the chief executives of the major organizations from where the interviews were drawn. Specific individual markers that might have identified various interviewees have been changed to maintain anonymity. One interviewee in a same-sex civil partnership was concerned that s/he would be identifiable, so that person's request not to mention the issue of sexuality in relation to other information has been honoured. Two of three of the interviewees with a disability asked that mention of their disability be precluded, and mention of disability throughout has been avoided. Certain features of ethnicity that may well have affected participants' home and working lives have also been removed, although some mention of people moving within Europe has been retained. Taking these decisions has limited the examination of the overlap between the barriers due to their gender and the possible perceived disadvantages due to the other factors such as sexuality, disability and/or ethnicity. Nevertheless, the requirement to maintain anonymity was paramount.

LITERATURE ON WOMEN IN SCIENCE

The literature referring to the lack of women in senior positions in science and why women mainly remain positioned in the lower grades is not huge and none refers to healthcare science. However, for the purpose of background and context, we now offer a short overview of what this literature includes. For example, Greenfield (2002a), Kirkup et al. (2010) and Lane (1999) pose the problem of why women do not gain 'top jobs' in science and discuss the perceptions of women working in science based on UK experience. Bilimoria and Liang (2012), Rosser (2004), Rosser and Taylor (2009), Valian (2000, 2004), Xie and Shauman (2003) and Zuckerman (1991) look at lack of progress of women in science based on US experience; Mallon et al. (2005) is based on UK and New Zealand scientists, Van den Brink et al. (2010) on the Dutch experience, and De Cheveigné (2009) on experiences in France. We look briefly at these authors in turn and then look at some of the other literature on women in science.

Kirkup et al. (2010), for the UK Resource Centre for Women (UKRC), provide statistics on numbers and proportions of women in science, engineering and technology (SET) occupations in the UK. Recommendations are provided for government and organizations, emphasizing the business advantages for promoting more women in science 'to boost economic growth' (p. 8):

Everyone involved in the career pipeline – government, education, business, and professional organizations – must build an 'integrated strategy'– seamless, systematic and coordinated, that takes account of modern realities: career paths are not necessarily linear or unbroken; education and employment must become more family friendly; gender stereotyping still influences subject and career choice especially for younger women; gender stereotyping throws up barriers in the workplace for employees of all ages.

Greenfield (2002a), in *SET FAIR: A Report on Women in Science, Engineering, and Technology*, covers many issues relevant to our research. The report lists possible reasons for the underrepresentation of women in science careers, many of which relate to 'informal practices' (p. 45) that result in gendered workplace cultures, gender imbalance including in the decision-making process and at appointment panels, and 'institutionalized sexism' (p. 28) but does not go into detail. Issues in education such as methods of teaching and 'stereotyping of careers advice' are also highlighted (pp. 30–31). The report also emphasizes the business advantages for the inclusion of more women in science.

Lane (1999) criticizes the lack of data collection in this country compared with the United States, which has been collecting data since 1981 on the numbers of women employed in science and engineering. She comments that 'mounting evidence and anecdotal reports tend to agree' that women face 'discrimination' in science due to 'gender bias' (p. 2).

Bilimoria and Liang (2012), assess the impact in 19 universities in the USA of the National Science Foundation's ADVANCE initiative for transformational change, which seeks to increase the representation and advancement of women in academic careers in science, technology, engineering and maths (STEM), promote gender equity and a more diverse science workforce, and has been running since 2001. The authors describe improvements in these areas to tackle the continuing underrepresentation of women and recommend ways to improve recruitment, advancement and retention of women.

Rosser (2004) similarly highlights the need for science institutes to change their culture rather than emphasizing how women need to change if they are to advance in science. In their 2009 paper, Rosser and Taylor (2009) ask why the problem of lack of women in science has yet to be resolved following 'years of debate and the investment of millions of federal and foundation dollars in programs encouraging women to enter science and engineering' (p. 7). Women achieve the same percentage of degrees as men but in science they do not reach the most senior positions in the STEM workforce. They discuss the barriers women meet, including childcare, lack of mentoring, and cultural and institutional biases, and make recommendations for change.

Xie and Shauman (2003) undertook a comprehensive longitudinal study in the USA comparing women's and men's careers to assess why women 'continue to be underrepresented in science' (p. 1). They show that women's advancement in science is influenced by personal issues and factors related to families and societal norms. Xie and Shauman (2003) argue that it is not marriage that hampers women's mobility and career progression, but having children: women are 'likely to forego their career goals altogether and replace them with family responsibilities' (p. 215). They found women's careers to be 'fluid' and changeable depending on circumstances (p. 209), arguing that 'causal mechanisms include socialization by parents, teachers, peers and media, role modelling, and perhaps overt practices of gender discrimination' (ibid.). Xie and Shauman (2003) also assert that women are less attracted to science careers than men and lack the aspiration to be scientists but acknowledge that the issues affecting women in science careers are multifaceted and complex.

Valian (2000, 2004) makes an important contribution in proposing 'gender schemas' that describe how subtle processes in science and academia act to disadvantage women:

> The main answer to the question why there are not more women at the top is that our gender schemas skew our perceptions and evaluations of men and women, causing us to overrate men and underrate women. Gender schemas affect our judgments of people's competence, ability and worth. (Valian, 2004, p. 208)

Zuckerman (1991) points out that much of the literature on the lack of progress of women in science has deficiencies, including being focused on academic scientists rather than industrial or government scientists, and this again highlights the gap in the literature that our book is helping to fill. Benschop and Brouns (2003) discuss how scientific quality can be 'constructed' differently in order to change the way universities (their research is in the Netherlands) tackle 'the slow progression of female researchers towards the top of academia' (p. 195). Although female scientists are not their main focus, their research highlights Valian's (2004) gender schemas illustrating that women scientists are undervalued are equally applicable elsewhere in academia. Mallon et al. (2005) focus on scientists undertaking R&D but do not use a gender perspective. Van den Brink et al. (2010) look at the processes of assessment in the appointments of professors in the Netherlands, noting that although formal practices may be in place in some universities, there are many gaps in these processes whereby the appointments can be manipulated to the detriment of women. De Cheveigné (2009) found in France that some women research scientists 'denied that they were actively discriminated against' (p. 128) whilst commenting upon individual cases of discrimination. Peterson (2010) shows in Sweden that women conform in technical settings to the masculine norm and understate their competence in order to be accepted in the workplace.

Much of the other literature on women in science gives historical perspectives, often focusing on the achievements of individual women in research science (see, for instance, Creese, 1998; Keller, 1983; Rossi, 1965; Sayre, 1975). Some literature addresses the medical issues that women face, where men take control of women's bodies and women are on the receiving end of medicine and technology (Barr and Birke, 1998; Ehrenreich and English, 1973, 1978; Greer, 1970, 1991).

A large body of feminist writings on science proposes that, as science is socially situated and political, it would be performed better if feminist and women's social values were part of the science agenda to give a broader perspective with the additional socially oriented contributions that women may bring (Fee, 1981; Harding, 1991; Schiebinger, 1989). Fee (1983) and others question such an approach and ask what additional contribution women can bring if science is truly objective.

Much literature highlights the masculinized nature of science and technology and the masculinized processes within science that subordinate women and remove choice (Burris, 1996; Flicker, 2003; Keller, 1985). Davis (2001) discusses the lack of women's networks in science. Some literature looks at women who are or are not attracted to science (Harding, 1991; Wajcman, 1991; Xie and Shauman, 2003). Phipps (2008) reviews UK initiatives on women in science since the 1970s and concludes that there is much to be done to improve women's underrepresentation in science.

There is limited research on non-medical women such as healthcare scientists working in public sector science, the focus of our book. Although not strictly addressing women in science, Witz (1992) examines gender and the power relationships and demarcation strategies established by the higher-status medical profession and healthcare professions such as radiography (which is highly relevant for the healthcare scientists in our research), and the role of gendering in nursing is explored by Davies (1983, 1995) and Davies and Rosser (1986).

A relatively recent book describes how the lives of women scientists are affected by having children (Monosson, 2008). Another (Evans and Grant, 2008) looks at how women study for a PhD when they have children. In addition, a wide body of literature is available on how women (although not women scientists) combine their roles as mother, homemaker, and wife with a paid job or career (Chandler, 1991; Charles and Kerr, 1988; Delamont, 2001a; Delphy and Leonard, 1992; Gatrell, 2005, 2008; Hochschild, 2003; Martin, 1984; Maushart, 2003; Potuchek, 1997; Williams, 2000). There are also authors who consider that women are less enthusiastic than men about undertaking paid work, and suggest that women work under sufferance and are not committed to careers (Hakim, 1995; Pinker, 2008).

Other authors also show that women continue to face and overcome more hurdles than men to make progress in their careers to overcome the disadvantage of their gender (Davidson and Cooper, 1992; Edwards and Wajcman, 2005; Höpfl and Hornby Atkinson, 2000) but science is not the focus of their literature.

Few authors explore in depth the barriers women in science need to overcome to progress their careers and this underlines the gap in the literature that we focus on in this book. Perhaps surprisingly, much literature from the 1970s remains relevant today. Science is changing rapidly, but women's place in science is moving much more slowly.

So, what is the role of government in assisting women in their science careers? Legislation in the form of the UK's Equality Act (2010) should protect women from discrimination but it seems that women are usually loath to invoke the law to counter subtle, albeit illegal, discriminatory actions at work (Gatrell, 2008). Reviews commissioned by recent governments (Greenfield, 2002a; House of Commons Science and Technology Committee [HoCSTC] report, 2014, quoted several times in this book; Organisation for Economic Co-operation and Development, [OECD] 2012) highlight many deficiencies in the education system, where they note that inadequacies in the education of women continue to disadvantage them in the workplace in comparison with men, even in the twenty-first century. In 2010, a report by the American Association of University Women (AAUW; Hill et al., 2010) similarly highlighted deficiencies in science education and careers in the USA where, '[t]he number of women in science and engineering is growing, yet men continue to outnumber women, especially in the upper levels of these professions' (p. xiv). This report notes the importance of culture in encouraging girls into science and highlights the role of negative stereotypes in reducing the performance of girls. Like the UK reports, recommendations are made, including changing the culture to encourage girls into science and engineering and taking measures to avoid bias against women.

In 1999, the UK Labour government set up the Athena project, which was originally funded as a four-year project and a further programme ran until 2007 with the aim to improve the progression of women in science. After 2007 the responsibilities were taken over by stakeholders including the UKRC and the Royal Society and more recently the Equality Challenge Unit (ECU) with its Athena SWAN Charter (see ECU website). The Athena SWAN Charter Award is administered by the ECU to research institutes who meet the required standards of practice:

In May 2015 the [Athena SWAN] charter was expanded to recognise work undertaken in arts, humanities, social sciences, business and law (AHSSBL), and in professional and support roles, and for trans staff and students. The charter now recognises work undertaken to address gender equality more broadly, and not just barriers to progression that affect women. (ECU, undated)

Whilst it is encouraging that other academic areas have now been raised in prominence by inclusion under the Athena SWAN umbrella, it could also be seen as disappointing that, as we show, progress on improving the situation for women in science has been limited and problems remain to be addressed. Unfortunately, applications to the Athena SWAN Charter are not possible in most of the institutes involved in our study, as the award is only available to academic institutes. Our research was mostly undertaken in public sector organizations, at least one of which was working towards Athena standards (ECU, 2016) even though it would not be able to receive the Athena award.

The Athena Forum, which has the same roots as Athena SWAN, was established as an independent committee in 2007–08. The Athena Forum works with UK universities, research organizations, professions and societies to address unequal gender representation across academic disciplines and professions, particularly the underrepresentation of women in academic science, technology, engineering, medicine and mathematics (STEMM) (Athena Forum, 2016). It is not involved with the Athena Swan Charter Award and has not expanded its remit more broadly.

In addition to government-led initiatives, over 200 websites are accessible and concerned with promoting women in science (Women-Related Web Sites in Science/Technology, 2017). The organizations variously aim to attract young women into science, help girls undertake work experience, encourage input from industry and advise organizations on good practice. A few of these such as Association for Women in Science (see AWIS website), Women in Science and Engineering (WISE) and Athena appear to have some influence on government policy.

Considering the increase in emphasis placed by successive governments on science and science education in schools since the 1960s, and more recently on women's careers, it might be expected that there would have been recognizable improvements in access for women to science careers. Sadly, there is still a long way to go to improve opportunities for women in science as we show in this book.

NOTE

1. For details on careers for biomedical scientists refer to the website of the Institute of
    Biomedical Science (IBMS); the current professional title is biomedical scientist but
    this has been changed twice over the years of Valerie's career. For details on clinical
    scientist careers refer to the websites of the Academy for Healthcare Science, Council
    of Healthcare Science, and NHS Careers. For details about the statutory registration of
    these two protected titles, refer to the HCPC website.

2. Positioning women in their place

[T]he dominant world-view is entirely from a male perspective, a perspective that has assigned to masculinity those characteristics which serve rationality, truth-seeking, logic. Woman, and the feminine, are cast in this scenario as the antithesis, the negation, and most particularly as the Other. (Haste, 1993, p. 5)

WOMEN'S PLACE

In this book, we use 'place' as a metaphor to illustrate how women in science are excluded from the 'place' of scientific 'action' (Miller, 1986, p. 75). Even though we are now in the twenty-first century, Miller's metaphor remains apposite to show how some men, both consciously and unconsciously, have over many years drawn on notions of 'difference' and 'otherness' to position women in lower grades. Here, we provide a taste of what this means theoretically, focusing on the UK and the USA.

Although 'place' may be interpreted in a variety of ways associated with geographical location, 'place' also has 'an extraordinary range of metaphorical meanings' (Harvey, 1993, p. 4). We argue that 'knowing their place' and the manner in which women 'internalize such notions psychologically' contributes to the consistent positioning of women at the margins, and lower levels within science (ibid.; see also Miller, 1986; Newman, 1995; Puwar, 2004). Women's 'place' in science and how far women progress (or not) within science careers assumes an established system of mutually understood hierarchies within scientific occupations. Our research interviews suggest that such hierarchies pervade the lives of women working in science, most of whom experience and reflect upon their 'place' at work on a daily basis. The women who participated in our study were keenly aware of the 'pecking order' among co-workers and were acutely conscious of their place and position within this order, reporting feelings of difference or 'otherness' regarding how they were placed in relation to male colleagues.

According to Harvey (1993), 'difference' and 'otherness' should be treated as 'something omnipresent' in an elaboration of space and place (pp. 3–4). We draw upon feminist debates about 'difference' and 'otherness' and suggest that both these concepts are utilized within science organizations to position women in a subordinate and disadvantageous 'place' compared with men. Women's 'place' in science separates them from the more 'influential' activities mainly undertaken by men (Miller, 1986, p. 75; Newman, 1995, p. 19; Puwar, 2004, p. 8). Arguably, despite over 40 years of equal opportunities legislation, in science the statement: 'It's a man's world' made by Janeway in 1971 remains relevant (p. 7).

The word 'place', like the word 'space', has connotations of separation and difference (Harvey, 1993). For Puwar (2004), 'space' refers to the reserved organizational 'places' where women and non-white workers are less welcome; their presence is noticed and incites opprobrium, as they are bodies 'out of place'. Puwar describes women and non-white workers who attain career roles more usually associated with men as 'space invaders' and a threat to the status quo (Puwar, 2004, p. 8). Sounding a cautiously positive note, however, Puwar notes that 'protected spaces can't be contained. They remain dynamic and open to other possibilities. Space is not a fixed entity' (Puwar, 2004, p. 1).

We also acknowledge in our analysis of interview texts, the importance of '[l]anguage [which contributes to] keeping women in their place' (Tannen, 1992, p. 241). The language of management, as of science, has been described as essentially masculine (Calás and Smircich, 1996; Collinson and Hearn, 1996; Tannen, 1992) and this causes difficulties for women seeking to alter and enhance their position within organizational settings. For example, conversational and managerial styles are interpreted differently depending on who is communicating with whom and, as Tannen (1992) observes: 'If a man appears forceful, logical, direct, masterful, and powerful, he enhances his value as a man. If a woman appears forceful, logical, direct, masterful, and powerful, she risks undercutting her value as a woman' (p. 241; see also Eagly and Carli, 2007).

Thus, women who attempt to adjust their styles by speaking more assertively may appear better to fit those models of masculinity that are associated with leadership and authority and it is possible that such women may command more respect at work. Equally likely, however, is the prospect that assertive women may be disliked and disparaged by co-workers as aggressive, pushy and unfeminine (Tannen, 1992, p. 239). So, as Tannen (2008) writes: 'Women in authority are subject to a double bind, a damned-if-you-do, damned-if-you-don't paradox. Society's expectations about how a woman should behave and how a person in authority should behave are at odds' (p. 126; see also Eagly and Carli, 2007).

FEMINIST PERSPECTIVES: DIFFERENCE AND GENDER

The values and qualities ascribed to women and men and how they are interpreted by society are based on perceptions of gender. In this section, we review the feminist perspectives that give rise to how differences are perceived and the effect they have on how people ascribe the value of a difference.

Feminist debates about 'difference' began with a liberal feminist discourse dating from the eighteenth century with Mary Wollstonecraft's *A Vindication of the Rights of Woman* (1792 [1999]), followed by John Stuart Mill's (1869 [2006]) 'The subjection of women', which influenced the suffragette movement. Liberal feminism became a force for emancipation that centred on notions of equality of access and opportunities (Schiebinger, 1999). In this view, it is hypothesized that if women were given the same access and opportunities as men, they would achieve equivalent career advancement. Fundamental to feminist liberal research agendas is the emphasis on gender stereotyping and sex segregation; they also highlight glass ceilings (or labyrinth[s]) that prevent women's progress to senior positions by blocking their way or sending them on circuitous routes where men tend to travel in a straight line (see Calás and Smircich, 1996; Eagly and Carli, 2007, p. 6). Cockburn (1991) describes equality of opportunities as 'breaking down barriers that prevent horizontal movement by women into non-traditional jobs, and removing those that confine women to the meanest jobs and prevent their vertical progress to different levels and locations in the hierarchy' (p. 46).

Though much criticized as being conceptualized around the situation of the privileged white middle class (arguably to the detriment of working class, ethnic minority and/or non-heterosexual women) liberal feminism has been the forerunner of much of the equal opportunities legislation in the UK and the USA. The reason for state alignment with a liberal feminist approach is possibly because this is regarded by governments as less threatening to capitalist economies (and as requiring less change from neoliberal agendas) than other forms of feminism might require (Gatrell and Swan, 2008) although Eisenstein (1984) argues that present legislation would not have taken place without the inclusion of more forceful feminist voices, particularly radical feminists. While legislation is far from being the complete answer, Acker (1998) suggests that little change will happen within organizations without it.

The UK's Equality Act (2010) states that there are nine protected characteristics to avoid discrimination and promote equality in the workplace: age; disability; gender reassignment; marriage and civil partnership;

pregnancy and maternity; race; religion or belief; sex; and sexual orientation (p. 4). Equality in these contexts identifies difference, and aims to treat
those who might be disadvantaged in such a way as to safeguard access to
the same opportunities as more privileged groups. In theory, as a consequence of equal opportunities legislation, evidence of disadvantage may
be challenged within a legal framework. However, the indications from the
women we interviewed who suffered discrimination were that legal challenges were avoided: as in Gatrell's (2008) research, women sought a career,
not an employment tribunal.

Within liberal feminist paradigms, the optimistic (though as yet
unproven) view has been that equal opportunities may reduce disadvantage
associated with difference and improve the situation of women, assuming
that, '[i]f the facts are known, people will change' (Delamont, 2003, p. 9).
Radical feminist perspectives locate gender differences as central and treat
'difference' as an 'alternative way of doing things, rather than as a deficit'
(Haste, 1993, p. 7; also see Eisenstein, 1984). Radical feminist perspectives
reject male forms of power and in their purest form advocate women-only
communities (see Calás and Smircich, 1996; Eisenstein, 1984). 'Woman-
centred feminism', where women are 'placed' at the centre of the debate,
is derived from the writings of lesbian feminists and illustrates the advantages for women if they are free from a commitment to men (Eisenstein,
1984, p. 48):

> By virtue of their lack of sexual ties to men, and therefore of their freedom from
> conventional heterosexual commitments, especially marriage, lesbians were
> placed sociologically in a situation of great freedom . . . Thus, they could think
> radically and profoundly without reference to gender arrangements. (Eisenstein,
> 1984, p. 50)

Radical feminism celebrates fundamental biological experiences shared
by many women (such as the potential for menstruation and childbirth),
and positions these differences as defining women's superiority compared
with men (Cronin, 1991; see also Calás and Smircich, 1996; Eisenstein,
1984; Haste, 1993). The emphasis on valuing feminine 'difference' is core
to many 'diversity' initiatives (Gatrell and Swan, 2008) and is emphasized
by Carol Gilligan (1982), who posits that women have essentially female
values that influence their ways of thinking, feeling and behaving, and are
undervalued by society's application of masculine standards and norms.

In contrast to radical feminism, socialist feminism minimizes differences
between the sexes almost to the point of androgyny (Eisenstein, 1984).
Socialist feminism arose out of challenging the continued inequality of
women compared with men despite nearly 100 years of women fighting for

equality. As such, it challenges the oppression of women and the power of men and contests the way women are constructed as subordinate to men in both public and private spheres. Oppression from capitalism originates in Marxist theories when work and social structure combine to subordinate working class men (Smith, 1974). According to Calás and Smircich (1996) socialist feminist theory brings together the best parts of Marxist theories (which do not adequately acknowledge women's situation) with psychoanalytic and radical feminism. Socialist feminism is thus closely linked with political and social action and considers that the oppression of women derives from both capitalism and patriarchy (Calás and Smircich, 1996; Eisenstein, 1984; Haste, 1993). Here, in accordance with a socialist feminist approach, we suggest that many differences between women and men relate to the social performance of gender and may be cultural (not biological) and that gendered behaviours are more likely be learned than inherent (Haste, 1993). As Wajcman (1998) observes: 'the values being ascribed to women originate in the historical subordination of women. The association of women with nurturance, warmth and intuition lies at the heart of traditional and oppressive conceptions of womanhood' (p. 60). In keeping with Wajcman's views, we argue here that women's lowered 'place' in science, compared with men, occurs as a consequence of organizational cultures and practices in science. A significant shift in attitudes towards women's position and intellectual value and potential is required within science if change is to be achieved.

Below, we argue that in science, men are positioned as dominant, with women positioned as 'different', not fitting into the masculinist ideal of creative scientific genius: men undertaking 'men's work' associate it with status and exclude women as 'other' (Davies and Thomas, 2002, p. 479; see also Hollway, 1996). Within science (and organizations more broadly), the theoretical 'differences' ascribed to women tend to be constructed as biological and inherent. Such supposed differences include a feminine aptitude for caring (partly due to women's capacity for reproduction) and apparently empathetic and intuitive personality traits (Haste, 1993; Gilligan, 1982; Noddings, 1984; Ruddick, 1989).

As a consequence, responsibilities to 'care for' (usually male) co-workers, and organizational systems often fall to women. 'Caring for' at work, across a variety of occupations, is often deemed to be 'low skilled' and is associated with 'low pay [and] insecurity' (Edwards and Wajcman, 2005, p. 38, citing Radin, 1996). In science, women's careers may stall as they are placed (and remain in) service roles supporting more senior men (Hewlett et al., 2008). This reflects inequalities of gender, race and class found in wider society where, '[w]omen find themselves either "serving" male bosses, or working in the "caring" industries in conditions that almost replicate their subordinate role at home' (Sims et al., 1993, p. 174; see also Hartmann, 1979).

Witz (1992) observes how in specialist settings (such as science, medicine and academia) women have historically been more likely than men to undertake 'semi-professional' roles involving support activities, effectively placing them in the 'role of handmaiden to male professionals' (p. 68; see also Fotaki, 2013; Ramsay and Letherby, 2006). Co-worker and line manager assumptions that care and empathy should come 'naturally' to women (Miller, 2005) are frequently invoked as a means of persuading female workers into such roles, even if this was not their original ambition (Ramsay and Letherby, 2006). Women in science are 'assigned', often by male line managers, tasks that men do not wish to undertake themselves, including caring and 'emotional work' (Harding, 1991, p. 47). Harding (1991) asserts that men are likely to work in 'caring' roles only if these are associated with prestige and command high fees (for example, psychiatry). It is well known that 'feminine' occupations concerned with caring and support (care worker, secretary) are poorly paid in comparison with roles more usually associated with production and masculinity (electrician, hospital consultant), perhaps due to associations with the unpaid work undertaken by women (especially mothers) in heterosexual households (Equalities Review, 2007). In relation to medicine, Code (1991) has illustrated how women's skills as a carer are discounted: despite medicine ostensibly being a caring profession, professionalization (especially in relation to top consultant jobs) often excludes women and pervades other branches of science:

> [W]omen's efforts to achieve authoritative status in the medical profession are both symptomatic and symbolic of the suppression of female knowledge and expertise. Medicine is a peculiarly salient example, for it enlists caring and nurturing skills long associated with women's essential nature. Yet the professionalization of medicine, together with its establishment on a scientific footing, produced an exclusionary structure in which experientially based female skills had no place. (Code, 1991, p. 227)

In particular, notions of genius and expertise are associated with knowledge and masculinities (Battersby, 1989; Code, 1991). As Code (1991) contends, women have suffered intellectually because they have historically been excluded from opportunities to become scientists.

In our research, we argue strongly that the supposed suitability of women workers to undertake caring and/or support roles is related to social constructions of gender, rather than any biological feature of womanhood. Yet social and organizational assumptions that women are innately possessed of empathy, and a resultant desire and talent for undertaking care and support work, affects the manner in which employed women are treated (we argue, especially in scientific environments; see Harding, 1991). This is because science is expected to be 'dispassionate, disinterested, impartial, concerned with abstract principles and rules' and such characteristics are associated with men (Harding, 1991, p. 47). Social assumptions about women's supposedly caring and intuitive characteristics inevitably exclude them from being perceived as ideal workers in a context where rationality is valorized as the ideal. And while women in science may be viewed as inherently lacking capacity for logical thought process, men are presumed to be naturally imbued with rational and systemic minds. Perceptions of women as intuitive but not necessarily rational have become infiltrated into organizational life to the advantage of men but to the detriment of women. As Founier and Keleman (2001, p. 268) argue:

> Organizational structures, cultures and everyday practices have all been shown to constitute the 'ideal employee' (and especially the ideal manager) as a disembodied and 'rational' figure, a figure that fits more closely with cultural images of masculinity than femininity (see Acker, 1990, 1992; Gherardi, 1995; Martin, 1989). Femininity, on the other hand, has tended to be associated with embodiment, emotions and sexuality; as such it is constituted as subordinate to 'male' rationality, and possibly as out of place in 'rational' organizations.

Hollway (1996) describes how men are positioned as dominant in the practices of management and science, and women (especially mothers) are positioned as different, as separate, as 'other' or, as we discuss in Chapter 6, perhaps more appropriately, as 'm[o]ther' (see Cooper, 1992; Fotaki, 2011). Hollway (1996) points out that men undertaking 'men's work' associate it with status and exclude women: 'Where work is defined as men's work in the gendered division of labour, we find that it relies on the other of women's work to invest it with masculine prowess or status, and thus on the exclusion or subordination of women' (p. 30).

As we show in the tables below, logic and rationality are thus associated with masculinity and favoured in the science workplace to men's advantage. 'Other' characteristics such as experience, intuition, serving and caring, are constructed by men (and some women) as apparently 'feminine' (see Bendl, 2008, citing Cullen, 1994; Haste, 1993). Such characteristics are welcomed so long as they keep women in their subordinate 'place' at work, arguably mirroring women's historic 'place' at home where women are responsible for domestic and care agendas, as Janeway (1971) depicts. However, in the science workplace, assumptions about women's aptitude for taking supporting (rather than leading) roles contribute to the maintenance of the status quo whereby women's 'place' is in the background. Tables 2.1 and 2.2 illustrate how values and qualities that are highly valued

Table 2.1   Values and qualities praised

Masculine	Feminine
High self-esteem	Reserved
Less respect for rules	Conventional
Autonomous	Conservative
Driven and single minded	Serving and caring
Confident	Reticent
Socially poised	Modest
Independent	Quiet
Charming	Faithful
Rational	Intuitive
Knowledge	Experience
Adventurous	Reliable
Risk taking	Cautious

Sources: See Bendl (2008, citing Cullen, 1994); Haste (1993).

Table 2.2   Interpretations of qualities

Masculine	Feminine
Firm	Stubborn
Decisive	Bossy
Intellectual	Frivolous
Ambitious	Aggressive
Stable	Unpredictable
Driven	Uncompromising
Successful	Pushy
Decision maker	Consultative
Logical	Emotional

Sources: See Bendl (2008, citing Cullen, 1994); Haste (1993).

in men are viewed more negatively when ascribed to women – traits associated with successful male leaders (or scientists) are perceived to be inappropriate and 'pushy' when exhibited by women workers.

WOMEN IN A 'SERVING' IMAGE

According to Prather (1971), one reason why women are attracted to certain roles is because they are frequently portrayed in a 'serving' image and women subconsciously imitate these roles: 'as serving others in the nurturing and caretaking roles such as mother, housewife, volunteer, or nurse. The origin of this image is the assumption that we innately, instinctively, or hormonally are adept at nurturing, sacrificing and caring for others' (pp. 16–17). Arguably, being portrayed in the image of servants in films perpetuates such images (Delamont, 2003). When women scientists are portrayed in popular fiction, they are rarely portrayed as leaders, as Flicker (2003) indicates: 'The role of the professional "scientist" is reserved for men; women are represented in less than a fifth of these roles' (Flicker, 2003, p. 316). Perhaps the lack of images of women as scientific leaders is unsurprising, given that few women were admitted onto science degree programmes prior to the 1960s (Dyhouse, 2006) and where they have succeeded in pursuing scientific careers, women are often written out of historical accounts:

> The notorious scarcity of intellectually authoritative women imposes a subtle contract on women's resistance to paternalism. Historically, there are notably few exemplary 'female knowers': no female Newtons, Descartes, or Darwins (though now we have Barbara McClintock). Hence women commonly have access only at second hand to ideals of cognitive authority. (Dyhouse, 2006, p. 187)

Whether or not fictional and historical portrayals of women as support workers (rather than as leading scientists and doctors) may be cited as a cause of limited opportunities, it remains the case that women are encouraged to take on caring and serving but rarely leading roles in science. This perpetuation of the status quo reinforces and reproduces the stereotypical positioning of women scientists as subordinate and inferior compared with male scientists.

Simone de Beauvoir indicated nearly 70 years ago in *The Second Sex* (1949 [1997]) that 'woman' is described with reference to 'man' as being the 'incidental, the inessential, as opposed to the essential' and quoting the French political and social philosopher and rationalist Benda, in his *Rapport d'Uriel*: 'He is the Subject, he is the Absolute – she is the Other' (Beauvoir, 1949 [1997], p. 16, quoting Benda, 1946).

It is important to recognize that serving roles, such as supporting junior staff and preparing for accreditation, are critical to the way a laboratory works. Despite such caring and serving roles being the responsibility of scientists alongside their scientific work, undertaking such work is not valued in the same way as creative research. Serving and caring roles take women away from the more high-profile science and, even when performed well, do little to recommend a woman as a serious scientist.

Ironically, as we demonstrate later in this book, the supposedly feminine traits of supporting, serving and caring are viewed in science as positive characteristics only when displayed by women who understand, and remain in, their 'place' at work. Should women attempt to step out of this place, positioning themselves, for example, as creative scientific geniuses (as in the case of Susan Greenfield), they may be treated as immodest trespassers, usurping the more prestigious 'place' in science more usually reserved for men. Our research interviews demonstrate how the association of genius and 'knowledge' with masculinity (Code, 1991) affects the perceptions and positions of both women and men in science.

WHY DO WOMEN REMAIN IN THEIR 'PLACE'?

Why is it that women remain in their 'place', or why is it, as Smith (1987) states, that '[w]omen are complicit in the social practices of their silence' (p. 34)? Kate Millett (1970 [2000], cited in Eisenstein, 1984), similarly asks 'how was it possible for patriarchy to continue in a world in which women had education, access to financial resources and extensive civil and political rights, and were not visibly subject to forms of direct coercion?' (p. 6).

While the feminist movement has consistently challenged gendered inequities at work, it remains the case that many employed women experience inequities on a daily basis, yet continue in their jobs. One of many explanations as to why women might continue working even when conditions are unfair, is Jean Baker Miller's (1986) theory of dominant and subordinate groups. In Miller's view, when dominant groups (for example, male healthcare scientists) hold power over supporting or subordinate groups (in our example, female healthcare scientists) this becomes the 'norm' for both parties. In everyday situations such as the workplace, overt challenges to this norm from subordinate groups may cause disruption and tension, as dominant groups resist any threat to their comfortable position. In turn, subordinate workers may avoid direct confrontation in order to circumvent conflict (knowing, perhaps, that the consequences for them as the weaker party in disagreements may be harsh they might develop indirect ways for achieving their aims; see Gatrell, 2005, for example, where it is observed how women treated unfairly at work may leave, transferring their skills to new employers who offer better terms).

Miller (1986) indicates that those who are subordinate within (especially gendered) relationships may have an acute understanding of how dominant groups will react to different situations and are 'able to predict their reactions of pleasure and displeasure' (p. 10). Arguably, such knowledge leads women in subordinate positions (both in science and other occupations) to accommodate the requirements of senior colleagues in order to minimize workplace battles.

Miller (1986) illustrates 'difference' in the behaviour of dominant and subordinate groups with regard to women and ethnic minority groups where the subordinate group is perceived by dominant group members as having certain characteristics 'more like children than adults – immaturity, weakness and helplessness' (p. 7). Dominant group members fail to recognize 'potential' in the subordinate group, who are excluded from manifesting characteristics that might challenge the dominant group (ibid.). Furthermore, 'Members of the dominant group do not understand why "they" [the subordinates] are so upset and angry' (Miller, 1986, p. 9).

According to Miller (1986), members of dominant groups consider those people in groups that they perceive as subordinate to be 'well-adjusted', but only provided they remain in their place. However, subordinate groups may be unfairly characterized as 'unusual' or even 'abnormal' if they seek to challenge the status quo. Hearn (1998) supports Miller's views and suggests that women and men are so used to living with gendered inequities that the opportunities for women to press for changes are limited.

Our interview data support these explanations for the consolidation of women's 'place' in the lower ranks of science. In the forthcoming chapters, we offer further explanations for women's endurance of their subordinate place, including in relation to the conflict with husbands/partners/children when women try to develop their careers.

We now provide short case studies to illustrate examples of two women geniuses and experts who have been marginalized compared with equivalent men in science and excluded from recognition: Rosalind Franklin in the 1950s and Susan Greenfield in the first decade of the twenty-first century. Both these women symbolize how women can be excluded from notions of genius and expertise.

Rosalind Franklin

Rosalind Franklin is an example of a significant exclusion of a female scientist from receiving due credit. Franklin was not recognized in her lifetime for her contribution to the discovery of the double helix of DNA, which gained a Nobel Prize for Crick, Watson and Wilkins in 1962. Although she died in 1958 from cancer at the early age of 37, she could have been awarded the Nobel Prize posthumously (but was not) as the rules excluding posthumous awards were introduced only in 1974. Franklin's contribution was, however, marginalized in the records of the discovery that merited the award (Piper, 1998; Sayre, 1975; Stent, 1980).

The account of Rosalind Franklin's exclusion from recognition was first recorded in 1968 in the book by Watson (1968 [1980]) when he foregrounded his and Crick's achievements. Maurice Wilkins (the third recipient of the Nobel Prize) is also often forgotten. Watson's account was critical of many contributors, of whom Franklin was just one. Aaron Klug, Franklin's last student, who received her papers after her death, wrote a paper to set the record straight, saying that Franklin was closer to discovering DNA than had been previously recorded (Klug, 1968 [1980]) but this account, like Franklin herself, has largely been ignored. In 1980, Stent (1980) republished Watson's book in a 'critical edition' together with various relevant papers including Klug's to present the discovery of DNA in context.

Rosalind Franklin had gone to work with Wilkins and clashed with him, apparently expecting to work as an equal but found that he expected her to work as his assistant. According to Piper (1998), Franklin's appointment was to conduct her own research in X-ray diffraction but although her position was confirmed in a letter of appointment, it was not made clear to Wilkins.

It seems that Watson considered that Franklin was invading the male 'place' occupied by Watson and his colleagues. Watson (1968 [1980]) makes several personal remarks in his book about Rosalind Franklin's clothes, her lack of lipstick and uses a familiarity in reporting by referring to her as 'Rosy', commenting: 'Momentarily I wondered how she would look if she took off her glasses and did something novel with her hair' (p. 45). Watson confesses to using some of her data without permission: 'By then it had been checked out with Rosy's precise measurements. Rosy of course, did not directly give us her data. For that matter, no-one at King's realized they were in our hands' (p. 105). One of the reviewers, André Lwoff (1968 [1980]), comments on this: 'It is a highly indirect gift, which might rather be considered a breach of faith' (p. 228).

Watson (1968 [1980]) records in an epilogue that, as Franklin had died, she had been unable to challenge the events described in his personal account. As though to make posthumous amends, Watson states, 'Since my initial impressions of her, both scientific and personal . . . were often wrong, I want to say something here about her achievements' (p. 132). He then accepts the difficulties of being a woman in science at that time: '[We both [Watson and Crick] came to appreciate greatly her personal honesty and generosity, realizing years too late the struggles that the intelligent woman faces to be accepted by a scientific world which often regards women as mere diversions from serious thinking' (p. 133). Although he precedes this statement with a paragraph about her achievements, he omits to praise her for her scientific ability. The epilogue is a somewhat poor recompense for the lack of recognition of her science.

Franklin has been recognized, not with a Nobel Prize, but with two blue plaques: one outside her flat in London and one on the wall at King's

College London 'commemorating all those who were involved in the DNA work: the four names – Rosalind's included – fit nicely in round the rim' (Piper, 1998, p. 154). She has also been recognized, but only since 2003, by the Royal Society, which now makes an annual 'Rosalind Franklin Award' of a medal and £30 000 to 'an individual [woman] for an outstanding contribution to any area of natural science, engineering or technology' (Royal Society, 2017).

Susan Greenfield

The second case study illustrating how a female creative genius and expert has been discounted is of Susan Greenfield, the Fullerian Professor of Physiology and Comparative Anatomy at Oxford University, who heads a team of scientists researching the genetics of Parkinson's and Alzheimer's diseases. She has been Director of the Royal Institution of Great Britain (RI), the first female to be its director in its 204-year-old history. She is a Baroness, a CBE and member of the House of Lords as one of the 'people's peers' appointed by then Prime Minister Tony Blair in 2000.

Susan Greenfield was born in 1950 in North London, the daughter of a Jewish father (an electrician) and non-Jewish mother (a dancer). She attended Godolphin and Latymer School, an independent school for girls, where she studied little science. Her family had little money and her father's brother paid her school fees (Franks, 2011). Greenfield studied philosophy and psychology at St Hilda's College in Oxford, obtained her doctorate in pharmacology, also at Oxford, and has many fellowships and honorary doctorates from universities worldwide. Together with this academic career, she has frequently appeared on television and has written several books with a popular appeal about the brain and its function. She is also a co-founder of two biotech companies that specialize in brain diseases. Although she describes herself as an atheist, she says that her Jewishness is important to her, describing herself as 'a secular Jew' (McCarthy and Spanner, 2000). Greenfield (2002b) has described in the press her own difficulties in her late twenties working in the masculinist world of science: 'It was about then I seem to recall, that I started feeling irritated, rather than simply thinking that it was part of my lot in life to have men – and it always was men – make comments such as, "You don't look like a scientist"'.

Susan Greenfield's more recent high-profile exclusion was not being elected a Fellow of the Royal Society. Tim Radford (2004) records in *The Guardian* that Susan Greenfield, although proposed for the long list of 535 nominations, was not one of 32 fellows of the Royal Society elected in a usually confidential process. This was confirmed by the Royal Society, who commented: 'The Council considers that the breach of confidence may have had the effect of damaging the professional reputation of Baroness Greenfield'. The Royal Society is renowned for electing very few women Fellows and in 2017 only 144 of the 1512 current elected Fellows were women (9.5 per cent) (Royal Society, 2017). Radford (2004) writes: 'the only name on the list of 535 original candidates to be revealed is Susan Greenfield's. Significantly, the only name now known to be not on the final shortlist is hers'. Radford (2004) comments: 'high-quality science is not the only reason for election. Fellowship is open to those who have raised public understanding or appreciation of science, or rendered conspicuous service in the cause of science'. The Royal Society took the unusual step of confirming that Susan Greenfield had been nominated, as Colin Blackstock wrote in *The Guardian*:

> A spokesman for the Royal Society refused to comment in detail on a story in today's Times Higher Education Supplement which says Lady Greenfield has been rejected, but told the Independent, 'The candidate in question was considered by the relevant sectional committees at the meetings in January and the decision was taken then that she, along with the majority of other candidates, would not be placed on the long-list for election this year. (Blackstock, 2004)

In follow-up articles, whilst some people supported Susan Greenfield's nomination, others went on record to say she would not be suitable. In a public statement Susan Greenfield responded: 'I do not understand how or why my nomination has been made public. I think it is a great pity that those who do not have the courage to identify themselves can make unsubstantiated criticisms both of my science and of my activities in public communication'. In 2017, Susan Greenfield remains excluded as a Fellow of the Royal Society (Royal Society, 2017).

Baroness Greenfield continues to attract controversy. She gave a speech in the House of Lords on the damaging implications of technology on neuroscience, and has since published a book *Mind Change* (2014), which warns of risks to the brain of overuse of the Internet and social media. Her views, however, have met with criticism from scientists and the media alike, saying that her views are not based on evidence and are misleading (see Bell et al., 2015; Chivers, 2014). Furthermore, following months of infighting, she lost her position as Director of Royal Institution following expensive renovations to the Royal Institution premises that she led on behalf of its Board. She threatened to sue on the grounds of discrimination but the case was settled out of court. The Royal Institute and Baroness Greenfield released a joint statement saying they had 'reached a full agreement as to the terms for Baroness Greenfield's departure from the post of director': 'The trustees also paid tribute to Greenfield's work at the RI and stated that "in light of recent press coverage, we wish to clarify that decisions regarding the recent refurbishment of the premises on Albemarle Street had support and approval of the governing council"' (Mair, 2010). The post of Director of the Royal Institution was abolished in 2010.

In conclusion, we propose (and will pursue this argument as we examine our data in Chapters 3–6), that women's place in science remains firmly set as secondary to that of equivalent men. While it is hoped that UK government initiatives such as Athena SWAN, and new 'game changing' programmes in the USA will effect change and enhance the situation of women in science (Hewlett et al., 2008), we argue that such plans may be challenging to implement while attitudes about women's 'place', in the second tier of science, continue to endure.

3. Subtle masculinities at work

[Most social studies of science fail] to take serious notice not only of the fact that science has been produced by a sub-set of the human race – that is, almost entirely by white, middle-class men – but also of the fact that it has evolved under the formative influence of a particular ideal of masculinity [associated with] 'virile' power.

(Keller, 1985, p. 7)

SUBTLE MASCULINITIES

In this chapter, we draw upon data from our interviews to explore the subtle masculinities that, we suggest, contribute to marginalizing women in the workplace that lead to women's underrepresentation in senior posts in science.[1] We seek to make visible, or by 'unmasking', (Hacking, 1998, p. 58) how women's role continues to be regarded by leading male scientists as that of homemaker and 'm[o]ther' (where women are expected to enact a gendered caring and serving role and which we discuss in detail in Chapter 6; see Cooper, 1992; Fotaki, 2011) while men's is that of 'breadwinnner' (Potuchek, 1997) and scientist, thus affecting men's views about women's work orientation and potential for career advancement. In Chapter 4 we suggest that similar subtle masculinities are also enacted at home.

Here, we show how subtle masculinities were enacted at work. In the first section, we show how male healthcare scientists subtly supported other men rather than women, followed by exploring how men specifically excluded women, possibly not consciously, but very effectively. This is followed by seeing how women undertake hidden work including relational activities and support work such as preparing the laboratory for accreditation visits while male scientists avoid these roles. As this hidden work is not recognized as being of scientific value, women are therefore denied access to opportunities in research science. Thereafter we look at so-called opportunities given to women and show how these may actually disadvantage a woman and serve mainly to benefit her boss. Last in this chapter, we endeavour to bring together the themes of subtle masculinities at work in 'Making connections'.

MEN SUPPORTING MEN

Science is a vital component of the world we live in and influences our everyday lives, driving technological as well as dynamic scientific advances, and is overwhelmingly dominated by men. Science has been a subject for feminist analyses over many years (Creager et al., 2001; Harding, 1991; Keller and Longino, 1996; Schiebinger, 1989), tackling how the subtle but damaging masculinist discourse affects the taken-for-granted 'patterns of behaviour, beliefs, symbols and identity reproduced' within organizations (Hearn, 2002, p. 42).[2] This masculinist discourse is the basis for the subtle sex discrimination that comprises 'the unequal and harmful treatment of women that is typically less visible [and] is often not noticed because most people have internalized subtle sexist behaviour as "normal", "natural", or customary' (Benokraitis and Feagin, 1995, p. 41). Because it is accepted as normal, both perpetrators and recipients may not be aware of how effectively 'subtle masculinities' perpetuate the positioning of women in a subordinate place in science.

It was clear from the interviews that women noticed how men supported other men in preference to supporting women, thereby excluding and disadvantaging female healthcare scientists. The interviews from the men confirmed this. Women considered how senior men were more likely to assist junior male healthcare scientists than women in equivalent positions. It appeared that junior men received more favourable treatment from male managers than did junior women. For instance, Beth reported that the boss's expectations of the men in her group were markedly different from what was expected of her. Beth's male colleagues did not have to follow the same management rules as she did and she felt that she was treated less fairly. She was in a small minority at her level and felt this inequality keenly:

> I knew I wanted to influence things but I've had to fight tooth and nail for everything I've got. The men seem to get away with murder in my opinion and if I was to say and do some of the things they do, then I don't think I'd be in the position I'm in. (Beth)

Consistent with Tannen (1992), who argues that working hard does not necessarily lead to success and career progression, Beth considered that she was treated less well than her male colleagues. She thought that there was a dual standard that her boss applied but did not notice, and she felt she had only made progress in her job by being compliant with his wishes. In addition she thought her male colleagues were not reprimanded, for instance, for being late for a meeting:

> He [Beth's boss] is very specific about time and if he's going to meet you he'll tell you I've got three minutes not five minutes not ten minutes but three minutes. If you're late for a meeting then you're in big trouble, me particularly, but one of the managers who is male, he can be an hour late and it's just laughed at because it's the norm for him. (Beth)

Beth's observations are in keeping with the findings of Kerfoot and Knights (1993), who note that it may be difficult for women to resist such unfair practices. The hegemony of male colleagues collectively protected by close male networks may compel women employees to comply with situations that appear unreasonable to them on the basis that resistance to deeply engrained, hegemonic patterns of behaviours might be career limiting.

Similar to Beth, Kate recalled the treatment of a male colleague, comparing it with her own treatment by their joint (male) scientist boss:

> 'Mike' [her manager] would never challenge 'Tom' [colleague on the same grade as Kate]. You know, I think it's because Tom's a man, but with me and other girls in the lab as well, I think, yeah, we all get challenged if there is a problem or something. But not Tom; no, not as much; not at all. (Kate)

This kind of story of men supporting other men rather than women was told by a number of the women including Jackie:

> 'Neil' [a colleague] is treated [advantageously] maybe [because] he is more intimidating than a girl. I don't know; I wouldn't say it's sexism except that, well, but there are underlying things you couldn't really classify as sexism. It's all these little subtle things and it is subtle you know. (Jackie)

Interestingly, Jackie, like Kate, referred to female healthcare scientists as 'girls', a term arguably reinforcing their subordinate position and status (Learmonth, 2009). Furthermore, in the previous quote, Kate refers to Tom as a 'man', again perhaps emphasizing that she considers he is regarded as having a higher status compared with her and the other 'girls', even though he is a colleague on the same grade and not her boss.

Angela, a senior healthcare scientist and head of a department, reported how her boss in a previous post favoured promoting men rather than women, but no one in her organization seemed to be concerned about it and it was 'just accepted':

> He was very supportive of men that he considered needed it. One example was a post-doc who he arranged an upgrade for when his wife got pregnant and had twins. No one did anything about it, though everyone knew. It was just accepted. (Angela)

Later, following reductions in budgets and consequent staff cuts, Angela's post became 'at risk' along with another female scientist:

> I had already told him that I wanted a new challenge. I think it was an easy option to make me redundant. The only other person made redundant was the other female scientist but both male scientists were kept on.

So Angela witnessed one example, and experienced another, of what could be seen as direct discrimination but that were subtly enacted and not challenged by the management hierarchy.

Beth, Kate, Jackie and Angela were all aware that the men they worked with received taken-for-granted cultural and organizational advantages performed in subtle ways. In accord with the views of Valian (2004), they observed that men were regarded more highly than they deserved but the women found it difficult to describe how and why the unfairness took place. These four women found it difficult to express their mistreatment in a similar way to one of Sonnert and Holton's (1995) female respondents, who said, 'I don't think I could have described well enough what I was experiencing when it was happening' (p. 129). Women in subordinate positions are likely to realize that they do not have the power to make changes and, as Miller (1986) describes, they prefer to keep a low profile to avoid a hostile reaction from their bosses. Jackie, for instance, became quite distressed as she spoke more about the overt discrimination she suffered:

> I don't know whether they are trying to keep me down or keep me in my place or keep me from progressing. Are they worried I'm going to leave or do they think I'm just not good enough at what I'm doing. Is that why? All these sorts of things go through my head because I see other people like 'Neil' [colleague] get on and they don't have to ask, whereas I have to fight to get training and stuff. (Jackie)

Jackie's concern about being kept in her place is consistent with Newman's (1995) observations to 'remember your real place', where women are expected to 'contribute' to the 'team' but are only valued to a limited extent (Newman 1995, p. 19). Jackie failed to negotiate a favourable position compared with her male colleague, which is in keeping with Babcock and Laschever's (2003) research that notes how women are generally not good at negotiating. Jackie found that her male colleague was treated differently from and better than she was, but not because of her inability to do the job. Neil was afforded preferential treatment and did not 'have to ask' to go on courses, whereas she had to 'fight to get training'.

Not only did women believe that men saw women as unsuitable for senior jobs compared with men (borne out in the interviews with men) but some of the men supported other men indirectly by constructing women as better suited for operational science jobs than men. Tim, for instance, a senior male scientist, offered a positive appraisal of women's potential as a follower, thereby positioning women firmly in their 'place' in supporting and less prestigious roles in science:

> I think women are much more focused and they're patient. If you give them a task they will do it; men will tend to deviate. And if someone comes along with something more interesting, men will move away from what they should be doing, and women don't. I mean we've got a study going on now where we've got four women working and they are so focused. You go in there and say I think we should do this, or there's a problem and we're going to try and solve it, try this, try that, and it's just done, you know they just do it. They come back with the results or they put it on the computer and the next thing you know you've got an e-mail with a pile of results on it. (Tim)

Tim's view of how a male scientist contributed was different:

> If you ask a man to do it he'll faff around for ages and then he might change what you asked him to do because he thinks it's better than what you suggested and so you have to go back and do it again. And there are things like that. And maybe I just think it's easier to work with women, there's less conflict. You do get more conflict working with men. They get all this peeing on the post sort of stuff. (Tim)

Tim's language apparently praised women but the praise was because women did as they were told. One example is the way he said he spoke to women: 'I think we should do this, or there's a problem and we're going to try and solve it, try this, try that'. Perhaps those on the receiving end of this way of communicating might feel they were being challenged, rather in the way Jackie described her manager challenging her, as quoted above. Similarly, although he did not admit it directly, Tim appeared to want to avoid conflict with his male staff, supporting the belief that male managers are intimidated by other men (again as Jackie commented above). Furthermore, Tim's remark about men 'peeing on the post' suggests that he also wanted to avoid any challenge to his status as the head of his laboratory. As a consequence, junior male healthcare scientists were allowed greater freedom to pursue their own research interests in a way that was not open to women.

Another aspect of male behaviour is apparent in this extract. A younger male scientist who thinks he might improve on his boss's scientific ideas might well disconcert his manager but he might also be deemed worthy of promotion. Conversely, the supposedly positive stereotype of a woman as cooperative and who does as she is told does not enhance her value in terms of advancement as a creative scientist: being acquiescent assists bosses but does not make her a credible candidate for a senior post. Furthermore, being helpful and obliging could well be interpreted as relational behaviour that 'gets disappeared' as Fletcher (2001) notably depicts: 'certain behaviours "get disappeared", not because they are ineffective but because they get associated with the feminine, relational or so-called softer side of organizational practice' (p. 3).

So women are condemned by some if they are too assertive and condemned by others if they are not assertive enough, in the double bind described by Tannen (2008): '[w]omen are subject to a double bind, a damned-if-you-do and damned-if-you-don't' dilemma (Tannen, 2008, p. 126). When women move outside the norm of the expected behaviour and the place associated with being a woman, they risk being seen as aggressive and are ostracized because of their difference. Women are expected to behave in ways befitting women and if they don't they suffer for it (Adler, 1993; Bendl, 2008; Eichler, 1980; Fotaki, 2011; Tannen, 2008). Tim was happy to have women around him to do his bidding and take his advice. However, the same women were allocated a lower place in the science hierarchy, discounted as potential senior scientists 'seen as offering feminine qualities only', thus excluding them from being accepted in the same way as a man (Davies and Thomas, 2002, p. 479; Wajcman, 1998, p. 77).

Some women found ways of coping with this double bind, even though they didn't like the way they felt obliged to behave. Jackie, for instance, was aware of what she could or could not say to her boss: 'I am very outspoken and I'm very verbal but I get on well with him because I'm submissive'. She realized that challenging her boss was counterproductive, so she coped by being 'submissive' and carrying out what he wanted. She realized she was being submissive and didn't like it but it was a way of ensuring that she got on with her boss. However, her submissiveness was also counterproductive as it seemed he interpreted her behaviour as relational, which perhaps he saw as being an ineffective way of working for a scientist (Carlson and Crawford, 2011; Fletcher, 2001). Jackie's acquiescence possibly contributed to her not being given the roles that she wanted that would assist her in her development as a healthcare scientist. Behaving in a way she thought meant she got 'on well with him' actually resulted in him disregarding her as a scientist and favouring her male colleague, Neil, as described above. Instead, Jackie's boss encouraged her to undertake support roles because she was good at them but these roles kept her in a subordinate position where she was regarded as a support worker.

MEN EXCLUDING WOMEN

A tendency for men to exclude women is longstanding, as Battersby (1989) comments: 'Like [De Beauvoir], I have argued that "Art", "Culture" and the whole of European "Civilisation" have been constructed via a process of excluding or denigrating female achievements' (p. 106).

With legislation such as the UK's Equality Act (2010), overt discrimination of women is often hidden. However, the tendency for men to support the career advancement of men with similar characteristics to themselves is well documented. Such subtle favouritism extended to men by male bosses accords with research by Ragins and Cotton (1999), who relate organizational facilitation of men's careers to effective sponsorship among and between men, while women experience poor access to informal networks. Informal, or 'old boy' networks (Reskin and McBrier, 2000, p. 227; see also Hersby et al., 2009) are associated with enhanced career progression and recruitment opportunities from which women are excluded. Thus, even in circumstances where appointment to new or higher-level roles is supposed be implemented via formal processes, line managers may manipulate such procedures in order to privilege colleagues from within their informal networks (Reskin and McBrier, 2000, p. 224).

Since women are often excluded from these networks, it is perhaps unsurprising that they are not welcomed into the science elite; in particular, they are excluded from decision-making forums, as a number of the women interviewed commented upon. Vanessa, a postdoctorate researcher in her late twenties, was one example:

> My boss is absolutely fantastic; he's lovely um but he's very close to the chap who was working on this project before and things tend to get decided between the two of them. (Vanessa)

Being excluded from managerial decisions did not help women escape from the subordinate place expected of them lower in the hierarchy. Margo was sure her sex was taken into account when decisions were made about her role but she did not have any opportunity to influence the decisions:

> When decisions are made here I think they do take gender into account. For example, there was one type of work when they thought, Oh! they've got to interact with burly men who will walk all over these females. They [managers] won't take us seriously so they give those kind of responsibilities to the men – I do seriously believe that. (Margo)

Often, such women felt that having scientific and managerial skills were not enough for them to be accepted into the inner circle and Margo saw as inequitable that the top layer of management in her organization was all men:

> We have our CEO [chief executive officer] and we have the executive committee at the top and they are all men. I get together with other team leaders for each section in other meetings but it's difficult to get the ear of the bosses on the executive committee. (Margo)

Similarly Alice, a senior female scientist, one level below her senior management team, commented on the significance of the absence of any women:

> I don't think the CEO is very good at involving women. It's strange, I thought I was comfortable with him, but he has surrounded himself by men. He is open in a lot of ways, but there is not a single woman in the group that is close to him. (Alice)

These accounts of being excluded are linked by a belief that men like to be 'close' to other men. It is clear that Alice sees the exclusively male membership of 'the group that is close to him' as a feature the chief executive deliberately chose. Furthermore, it was not because women were unavailable; rather, she believed the chief executive had difficulty relating to women on a professional level. As Alice further explained:

> They are all men at the top and I don't think the CEO includes women. I think women are very good managers and support their staff but from what I have seen, not many get up to the top. (Alice)

Here, Alice expresses support for her female colleagues, which is a different observation from Rhoton (2011), who records that senior women in her study did not support other women. Alice viewed good management as requiring a relational style with staff, something that the women in Rhoton's study possibly rejected as being feminine and ineffective (Carlson and Crawford, 2011; Fletcher, 2001), perpetuating a masculine style of management. Alice considered that men were promoted based on their ability to demonstrate confidence, rather than for their abilities to be good scientists, leaders and managers. One of the most obvious consequences of men's lack of recognition of the value of a relational style of management was in recruitment decisions that led to further exclusion of women:

> I'm generalizing, but on the whole, from what I've seen, men are able to talk a lot and to appear very confident and that seems to get them promotion, but sometimes [there is] little to back it up. One good example is 'Jim Brown'; he has just been promoted to the same level as me because he's a very confident

person. You'd think that you'd have to have good management skills, good
team-building skills, as well as being a good scientist, which would be valued
highly, but not here. (Alice)

Ben, at 34, a relatively young postdoctorate scientist, put his success at
being interviewed down to his own confidence:

I get the impression that when I'm interviewed I come across well. I'm quite
confident and people respond to that. I've not been to many interviews so my
ratio of getting jobs and not getting jobs is not bad. On paper, you know, I'm not
a superstar academic or student. I didn't get grade 'As'. OK, I got a first degree
but not from a top university and I have not published that many papers. (Ben)

Thus, an ability 'to talk a lot and to appear very confident' as Alice suggested and Ben demonstrated, were the supposedly 'masculine' characteristics on which decisions about appointments were made. Such 'masculine'
characteristics were apparently regarded as being more important than
the official criteria specified for appointments, which Alice pointed out
included 'good management skills, good team building skills . . . being a
good scientist'.

Alice's observations were substantiated, perhaps not surprisingly, by
the three very senior men interviewed, who each expounded the ways they
assessed people for promotion (see Adler, 1993; Schein and Davidson,
1993). Frank (CEO [chief executive officer] and chair of a board), Ralph
(CEO) and Liam (CEO and board member) ostensibly regarded women as
being appropriate for senior appointments in their organizations, at least
in theory. In practice, however, men were appointed to leadership positions
and women were not. Whilst using the rhetoric of inclusion, they perpetu-
ated the exclusion of women from the place of men, the place of action
apparently due to women's lack of confidence, so keeping them in the place
of women (Miller, 1986, p. 75):

In some ways women are held back because they will not perform at interview;
they are not fully convinced that they should be there. I've been amazed at some
of the CVs, then I look at the person and talk to them and they do not have the
confidence commensurate. (Liam)

So Liam, in a different institute from Alice (but as she had also found in
her own place of work), admitted excluding women because he based his
appointments on confidence, which he apparently regarded as a more
important criterion than having amazing CVs.

Frank, a CEO of a public sector science institute, commented on the
constitution of the all-male board he chaired where no women had been
appointed:

We're a small board, but it's not, it's certainly not deliberate policy. There are a number of potential candidates for board level among the female community; I mean I generally don't consider it an issue, but I'm very conscious of the fact that people look at us and say, 'They are all men'. (Frank)

Although Frank said he noticed that people had commented upon his all-male board and asserted that he didn't consider it an issue, it appeared he was ill at ease with disclosing it. He might also have realized that the 'people [who] look at us and say "They are all men"' were not able to influence the board's composition – that would have been his decision. What is also remarkable in view of the Alice's comments about her CEO's failure to involve women is the phrase 'the female community'. By using it here, Frank seemed to concur with Miller's (1986) view of women being in a separate group and one from which his interests, values and emotions were distinct and detached from his own views. Furthermore, in alluding to a separate 'female community', he implied that this female community was inferior. Although Frank's board might have been described as a male community, he did not comment on it as such.

Frank's depiction of the 'female community' is an example of the way in which the subtle masculinities are enacted within science. The effect of such subtle masculinities reinforces women's position by emphasizing women's difference and 'otherness' and is in accord with the comments by Hearn et al. (2009) with regard to management: '[m]anagement overwhelmingly remains men's arenas . . . with clear structural gendered hierarchies' (p. 42). In such a way, women's exclusion from Miller's (1986) place of the action becomes accepted and endorsed. In accordance with Bendl's (2008) description of the 'gendered subtext', Frank highlights the dichotomy of 'males as the norm' and 'women as the other' (p. S56). Perhaps Frank used a language of inclusion 'in order to make the board appear gender neutral' (Bendl, 2008, p. S57, citing Bradshaw, 1996) and to imply support for the advancement of women whilst simultaneously endorsing an agenda of exclusion. As Hearn (1998) points out, the 'taken-for-grantedness of men is reaffirmed through the absence of men. Men are unspoken and so reaffirmed' (p. 787).

Writing in the context of the advancement of managerial executives, Kanter (1977, pp. 62–3) uses the term 'homosocial reproduction' to describe how men promote other men rather than women:

Keeping [top] positions in the hands of people of one's [own] kind provides reinforcement for the belief that people like oneself actually deserve to have such authority. 'Homosocial' . . . reproduction provide[s] an important form of reassurance in the face of uncertainty about performance measurement in high-reward, high-prestige positions. So [top] positions . . . become easily closed to people who are 'different'.

That this subtle, possibly unconscious, gender bias remains prevalent is confirmed in the House of Commons Science and Technology Committee (HoCSTC) report, *Women in Scientific Careers* (2014), which cites evidence submitted by the British Medical Association and Newcastle University: 'unconscious bias means that panels frequently have a tendency to choose appointees like themselves'. This type of bias in an environment dominated at senior levels by men may mean that 'many successful candidates will be male' (p. 19).

As Fotaki (2011) points out, to succeed in a man's world, women must comply with the 'limitations of their embodied presence – their comportment, expression, speech and so on' (p. 48). Liam saw lack of 'confidence' as a primary reason for not recruiting senior female staff and if women did not show sufficient confidence to meet his personal criteria they were not appointed. It needs to be remembered, however, referring to our tables in Chapter 2, that while confidence may be valued in a male colleague, women displaying similar behaviour would not necessarily be welcomed (see Bendl, 2008, citing Cullen, 1994; Haste 1993). Arguably then, if women in our study had shown confidence, they might still not be appointed because their confidence might mean they were labelled as aggressive or pushy.

Furthermore, it seemed that male appointment panels could manipulate human resource advice on interviewing so that women were apparently excluded because they were classified by, in this case, the all-male appointment panels, as lacking 'merit and competence', as Frank related:

> We had three women apply for the most recent recruitments to the board, but they didn't get it as the promotion appointments are still based on merit and competence. (Frank)

When encouraged to say more about what constitutes the nature of criteria like 'merit and competence', Frank attempted to describe the impartiality of the appointment procedure but admitted to the subjective element:

> I think most people in [his organization] do try to say when they are doing appointment panels, 'What are the competencies required for this job?' and 'Who fulfils those competencies best?' although I always think it's difficult to remove the subjective. (Frank)

By saying 'I think most people' and recognizing the 'subjective' element in his organization's appointments, Frank was seemingly unsure whether people in his organization conducted fair appointment panels or otherwise, which was quite an admission from someone who was the overall boss. In denial, Frank then said he did not consider the lack of women at director and board level as 'an issue' because there were women being promoted to the level below:

> Certainly during my time here more and more women come through to head of department level. I mean I generally don't consider it an issue. (Frank)

Interestingly, one of Frank's female heads of department (who was also interviewed for this research at Frank's suggestion) put a somewhat different slant on the numbers of potential female candidates available for board-level appointments in Frank's organization where he was CEO:

> The directors are all men so the next level down are the heads of department and there's me, Barbara, Lisa and Daisy – four women out of about 15 – not very many is it?

When invited to talk about the interview process, Ralph, a CEO in another organization, made particularly enlightening comments, depicting different characteristics to be more important:

> We interviewed for one job, and there was a serious internal candidate, and the reason she didn't get it was because people regarded her as too aggressive and wouldn't make it work; [she was] very able, but divisive. (Ralph)

So this internal woman candidate was condemned and excluded before she started the appointment interview. In noting Ralph's opinion that the woman was 'very able', but her ability was outweighed by being considered 'too aggressive', it could be questioned what counted as 'too aggressive'. Referring again to our tables in Chapter 2, and recalling the views of Bendl (2008) and Tannen (2008), it could be argued that '[i]f a woman speaks or acts in ways . . . expected of a woman, she will be liked but may be underestimated. If she acts in ways . . . expected of a person in authority, she may be respected but will probably be viewed as too aggressive' (Tannen, 2008, p. 127). Ralph believed he 'knew' the woman to be aggressive because he had insider knowledge; the other candidates could have been equally or more aggressive but he wouldn't have known that about external candidates. Furthermore, had a man demonstrated such traits, he might have been applauded for his decisive and determined approach. Valerie then asked Ralph if he had counselled the woman about her apparent 'aggressiveness', to which he replied, 'I must remember to do that'.

As highlighted in Chapter 2, what was considered by Ralph as aggressiveness in a woman could well be interpreted in a man as someone who was ambitious or displaying confidence, and be welcomed (as Liam considered) as a positive characteristic. This dichotomy of language also illustrates the double standard by which women live where 'identical behaviours are not defined as the same but as different due to the sex of the performer and the social context in which they take place' (Eichler, 1980, p. 16). Furthermore, as Eichler identifies, 'our language is tied to sex, our vocabulary is not adequate to describe the absence of sex differences when it occurs' (Eichler, 1980, p. 14). Women professionals are thus faced with what Laws (1975) terms a 'double deviance' where they have seemingly two factors of difference – not being a man and being so-called aggressive.

Women who demonstrate qualities that are associated with masculinity and that may be valued among male colleagues (drive, personal ambition, risk taking and self-confidence) may be seen as inappropriately masculine (Eagly and Karau, 2002; Schein, 2001). Women scientists who conduct themselves at work in a style that is regarded as unfeminine are likely to be disadvantaged because they are seen as pushy and aggressive, while those who are helpful, quiet and supportive may be dismissed as unsuitable for promotion due to their apparent lack of drive and confidence. Yet at the same time, behaviour associated with the stereotype of femininity such as being caring and supportive of others, deference and careful management of resources, which are considered to be attractive among women workers, are rarely regarded as criteria for promotion (Eagly and Karau, 2002; Eagly et al., 2003).

Both men and women are expected to 'perform' in their roles in expected ways and to 'fit in'. On the one hand, women are expected to conform to a dress code as men do, but for women the expectation is to appear to be indistinguishable from men, as Wajcman (1998) indicates: 'women who have made it to senior positions are in most respects indistinguishable from the men in equivalent positions' (pp. 55–6). Perhaps this 'role trap' of 'iron maiden' (Wajcman, 1998, p. 110) is a survival strategy for women and a way of coping with the characteristics expected of someone who aspires to get to the top.

It is not surprising then, that Angela, head of a department and a senior healthcare scientist, commented that she had to behave carefully with a boss who had a different set of values:

> I went to one meeting and tried to contribute some ideas. But I find there isn't a safe environment where I can express my ideas and not get shouted down and made to feel as though I've said something ludicrous. I wasn't invited to those meetings again. It was probably the wrong way to contribute to the discussion but I give loyalty to my boss and follow as well as I can. I think I'm

truthful and honest and I expect everybody to play by the rules that I've got but most of the time the men abide by one set of rules and we have another. (Angela)

Angela echoed Fotaki's comments on the 'unrepresented woman' where the 'masculine logos [is] the dominant form of speech' and where women may be 'demolished' in a public forum where 'would be participants may ... be silenced' (2011, pp. 46–7). Angela was frustrated that she could not contribute as an equal and remained excluded. Furthermore, it appeared that she felt that she was the one at fault, noticing her difference and assuming her deficit.

Alice, also a senior healthcare scientist and head of a department in another organization, similarly commented on her lack of respect for her male boss, saying he was poor at communicating and failed to involve her. Her relational way of working was important to her too:

> You have to bring your team with you – I rely on my staff to help me and I think it's a two-way thing and they know that I will support them. (Alice)

Alice's working approach was in contrast to her boss's, who she felt excluded her from his decision making:

> I don't think I've learned anything from my own head of department. He is not good at communicating and he does not work in the same way as I do. I see decisions made behind closed doors. I'm probably not one of his favourites but I'm good at my job so he can't get rid of me. (Alice)

Her exclusion is in line with the proposal from Carlson and Crawford (2011) that women are excluded because they act in a relational way that may be viewed by senior male scientists as ineffective. This lack of recognition for skills other than strictly scientific skills was also recognized in the HoCSTC report as disadvantaging women:

> Evaluation of success in STEM jobs typically relies heavily on 'quantity' ... technical ability and intellectual rigor, but often fails to formally highlight and recognise facets of ability which have a significant impact on actual performance. For example, academic scientists spend a considerable proportion of their time communicating (in articles, at conferences and seminars), networking, writing grant proposals, supervising students, managing staff, teaching and – increasingly – performing public outreach activities and working on the commercial exploitation of their findings. (HoCSTC, 2014, p. 22, quoting evidence from the Science and Technology Facilities Council, Women in Science, Technology, Engineering and Maths Network [STFC WiSTEM])

Alice thought that she acted with integrity and inclusiveness but her male boss excluded her; her phrase about her boss 'not work[ing] in the same way as I do' possibly indicated his rejection of her difference. Furthermore, her relational way of working with her own staff was in direct opposition to his way of working with her, and we interpret this as an unspoken explanation for why he excluded her from discussions that affected her and her team. Her bitterness at the way her boss treated her is expressed by her believing that he would prefer it if she did not report to him.

Another example was Carla, who reflected on the situation described by Adler (1993) where women and men are expected to dress according to their gender and their role, and indicated that she behaved in the particular way expected and accepted of people in 'public places' (Goffman, 1963). Unlike the situations that Puwar (2004) describes where women and minority groups are noticed as outsiders, Carla fitted her identity to her surroundings:

> I don't think it's being a fake person; I think it's all part of yourself. I went into a meeting the other day and all these guys were there, all the same age all dressed exactly the same and in the same outfit. Just for a minute I thought, Oh, I should have worn a tie. (Carla)

These observations echo Coffey (1999), who writes that she wore a dark suit and painful high-heeled shoes to conduct research interviews to blend into the environment. When Carla was interviewed for our research she wore a black trouser suit with a white shirt. As she commented, she wore no tie but nevertheless would have blended in well at a meeting of besuited males.

Paradoxically, women are also expected to perform their femininity (Wajcman, 1998). Marjorie Scardino (who was the first woman chair of a FTSE 100 corporate board) comments that 'many of the decisions about who goes on [the board] are determined by the chemistry of the new prospective board member and the board itself' (quoted in Vinnicombe and Bank, 2003, p. 66). Vinnicombe and Bank (2003) report that in Scardino's experience 'women could do more to make themselves compatible with boardroom behaviours and environment', telling them that she knew of two women who had been rejected because the board chair thought the women were going to be 'too serious and in a way too aggressive' (Scardino, quoted in Vinnicombe and Bank, 2003, pp. 66–7). In other words, Scardino acknowledges that the board would appoint people in its own image, in what Kanter (1977) terms 'homosocial reproduction', but excluding women who manifested some of the characteristic associated with men rather than women, so perpetuating the status quo (p. 63).

Scardino implies that such behaviour by board members is acceptable, and that women need to change their behaviour, echoing the woman as the 'deficient' model (Miller, 1986; Rosser, 2004; Wajcman, 1998). There is no criticism of board members who might need to change to accept women into their protected space. Scardino does not appear to comment (in public, at least), that such behaviour could be seen as discriminatory.

Some men were openly direct about their exclusion of women and rationalized their behaviour. Part way through his interview, James was invited to consider that we exist in a world where the values we live by are derived from masculine standards. James was adamant that we do not. James considered that women competed with men on an 'equal playing field' in terms of accessing opportunities in the world of science and was sure that he was not biased. He also considered that those women who did not progress in their careers, either 'failed' because they were not good enough compared with men or alternatively, and consistent with Perry (1992), because they 'chose' not to progress:

> As far as it is I'm concerned it's an equal playing field. Women get just the same opportunities as men and are treated just the same. It's up to them. If they're good they'll get on but at the moment there aren't enough good women. (James)

In this example of 'distorted' thinking described by Kanter (1977, p. 211), James made generalizations indicating his lack of knowledge about the gender imbalance in science and about legislation and government initiatives (Equality Act, 2010; Greenfield, 2002a, 2002b) to try to correct the gender inequities in science. He also indicated a lack of interest in exploring the topic further or, unsurprisingly, of righting any wrongs in his own organization. When discussing recruitment practices, James reflected:

> I'm a scientist and as a scientist, I'm objective. I don't find it at all difficult to make judgements about the scientists I'm interviewing. I don't need all this diversity training. It's a load of twaddle. (James)

James's own progress had been smooth, opportunities had been offered and he had not met barriers to block his progress. He considered, 'If I can do it, anyone can do it', having no understanding that a woman in a similar position might find insurmountable barriers blocking her path and keeping her in her place. He demonstrated what we term 'discrimination blindness', seeing women's different career needs as 'other' and alien to his comprehension:

> It really annoys me that people suggest we aren't fair in our selection procedures. Why should women get special treatment? They have the same chances as the

rest of us. There are loads of women in medicine and hospital management now. (James)

Thus, James reinforced structural determinants (Kerckhoff, 1995; Maranda and Comeau, 2000) and gendered professional discriminatory practices (Witz, 1992) within his own organization, which hindered women's progression. Like most health organizations in the public sector, there are more women in the lower grades but men dominate in the higher grades. James 'distorted' the situation in his organization, using examples to support his case from outside his sphere of influence, referring to the increasing numbers of women in hospital management and women doctors to illustrate his point and to absolve his own responsibility (after Kanter, 1977, p. 211). It had apparently never occurred to him that lack of women's progression might be due to reasons other than having 'equal opportunities', a rhetoric that needs to be translated into action, but to which James was resistant.

James also seemed deficient in the relational skills (Fletcher, 2001) that might have helped him understand the needs of his female staff for career progression (p. 14) and he disregarded his own responsibilities in developing the women who worked at the more basic grades in his laboratory. James accepted the masculinist discourse that assumed women not only had the same opportunities as men but were also able to access and benefit from them; in doing so James maintained power in the place of the action for men and excluded women (Miller, 1986).

Will, a senior scientific manager in a public sector organization leading a specialist scientific team of about 20 people, believed that men are more suitable than women to head laboratories because he said, there were more men in his organization:

> Our organization has always been numerically top heavy with males, so I suppose it's more likely to be a man who'll rise to head laboratories or run teams. The ratio has always been more men. It's just the history of how we grew up. (Will)

In Will's public sector organization, women represented over half the staff but this did not alter his 'distorted' (Kanter, 1977) perception that it was 'top heavy with males', perhaps because his assessment was influenced by seeing more men than women at the top of the organization (p. 211).

Will's expectation was that men would head laboratories in his organization. He seemed content to accept the status quo, seemingly seeing women as separate and 'other' and not embodied as suitable to be leaders, emphasizing his expectations of a stereotypical male head of a laboratory (see Brewis, 1999; Davies and Thomas, 2002, p. 479; Hollway, 1996). Consistent with Miller (1986), management in science was, to Will, men's place of 'action' – women were positioned as subordinate, invisible, and he ignored and excluded them, apparently not feeling that he needed to be overly concerned about them (p. 75). When Will was asked whether his organization made any conscious effort to redress the uneven numbers of women at the top of his institute, he said he thought that this had never happened:

> I've never met that attitude at all either for or against really – pretty neutral. (Will)

Although Will said that the 'attitude' was 'pretty neutral', he used the word 'attitude' in an unusual context. 'Attitude' is usually understood to mean 'standpoint' or 'way of behaving' and Will implied that it would be anomalous or aberrant behaviour if his institute had taken a stand towards encouraging more women to be promoted to the senior management of his organization. Will believed that no one in his organization had ever been 'for or against' increasing the numbers of women at the top of his organization – it had never even been considered.

Another area where men excluded women was in negotiations, particularly over pay: women rarely negotiated over pay or promotion. Most of the women interviewed were also unsuccessful in negotiations with bosses for access to the education and training they needed to apply for higher positions, especially for access to study for PhD, the near obligatory initial passport on the way to the top in science. Research by Bowles et al. (2005) and Babcock and Laschever (2003) indicates that women negotiate far less than men and when women do negotiate they ask for less, and get less, than men: 'Conventional wisdom (such as "it pays to ask" and "the squeaky wheel gets the grease") suggests that, if women want the same resources and opportunities as men, then they should learn to seek out, rather than shy away from, opportunities to negotiate' (Bowles et al., 2005, p. 3). However, these authors show that when women negotiate they are viewed as confrontational – negotiation is not something that is expected of or accepted in women and it is counterproductive for women to negotiate in ways perceived as masculine. Women who put themselves forward in a way perceived by others to be masculine are less likely to be appointed and overall are judged more severely than men (Bowles et al., 2005; Rudman, 1998). Bowles et al. (2005) discuss the difficulty women have in negotiations on pay both at the appointment stage and later when negotiating a pay rise or a better position for themselves, where they are disadvantaged because of the typecasting of others and of themselves.

Beth, who worked in the commercial sector, was the only one of the women interviewed who said she challenged her boss over money:

I had a major op last year and was off work for ten months. My boss said he couldn't believe I could take holiday after being off sick and when my annual salary increase was given to me, I was told that I wasn't entitled to the full amount because I'd had a lot of time off sick and hadn't done a full year, so I was absolutely furious with him. I didn't know what to say and went away and thought about it. Then I came back and told him I had taken it higher, to the CEO. They have corrected it to some degree but it's too late: it's not the money, it's the principle. They have said they've found it very hard to manage without me and they realized how much I did and I thought, well all these are very good positive signs and then to turn round in the next breath and say well we can't give you the full pay rise – unbelievable. (Beth)

The idea that women should not 'ask' remains an issue, as Beth discovered. As a result of being denied the money she was entitled to, Beth disengaged from her organization. On the one hand, her company said how much they valued her but did not follow this up by paying her what she was due. On the other hand, they perhaps thought they could get away without paying her what she was due because it was unlikely that she would challenge them. She was also apparently being punished for having an illness only suffered by women.

On a global scale, at a conference in October 2014 to celebrate women in technology, the newly appointed chief executive of Microsoft, Satya Nadella, suggested that women not asking for a pay rise was 'good karma'. His comments invoked anger among staff and the public alike. In a subsequent email to staff, Nadella apologized, saying he 'wholeheartedly' supported programmes to close the pay gap (BBC News, 2014). Yet Nadella's observations about women's place within negotiating arenas are in keeping with the experiences of many women attempting to enhance their pay and position in organizations. Sheryl Sandberg (2014), chief operating officer of Facebook, relates a number of uncomfortable conversations experienced by very senior women (including herself) who are required to be tough negotiators as part of their jobs but who are not invited to request better terms and conditions for themselves. Sandberg observes how (often male) line managers may be notably unenthusiastic about women's questioning regarding terms and conditions of employment, in a manner that does not appear to apply to men in equivalent positions. Given this it is perhaps unsurprising that gender pay gaps between women and men persist, with women's pay lagging behind that of equivalent men despite equal pay laws having been in place since the 1970s (Blau and Lawrence, 2000).

We now look at another aspect of men's behaviour that arose in the women's accounts: that men avoided or were not asked to take on support and relational roles, which we view as hidden work. Because women tend to take on or are allocated such support and relational roles, like Jackie, they were not seen as being potential high-flying scientists and were kept in women's subordinate place.

HIDDEN WORK

While working in science is not directly related to caring or nurturing, for most of the women interviewed, their roles at work seemed to satisfy their need to do 'something good', using language like 'making a difference', 'doing good' or 'helping' patients. For instance, women acted as informal counsellors, dealing with people's troubles, often in quiet, out-of-the-way places, sometimes making themselves available out of hours to help more junior staff with training issues or providing other advice. In line with Fletcher (2001), work that relates to relational activities that women do well where they support other people is not recognized as 'real' work and becomes 'disappeared' (p. 3). Similarly, the HoCSTC report (2014) notes how these relational activities, termed 'soft responsibilities', are not highly valued:

> Interestingly, the skills that are normally considered essential to leadership are under-valued in academia: *ScienceGrrl s*tated that 'non-research skills (for example, leadership, mentoring, pastoral care, teaching, project/lab management) appear to be largely ignored' in career advancement. This can be a gender issue as 'anecdotally . . . more women than men take on so-called "soft" responsibilities'. (HoCSTC, 2014, p. 22, quoting *ScienceGrrl*)

In addition to these kinds of relational activities, the women in our research tended to find themselves responsible for hidden work that men also avoided, such as the activities needed to keep the laboratory functioning, including preparing for accreditation, checking that equipment was being properly monitored or even that the laboratory was kept tidy (necessary to conform with health and safety requirements in laboratories). These duties inadvertently labelled women as the metaphorical 'm[o]ther' and support worker and so disadvantaged them by associating them with the place of home (as discussed further in Chapter 6; see Cooper, 1992; Fotaki, 2011), conforming with gendered expectations. Women also took on management responsibilities that were less highly regarded than scientific research and development (R&D) by scientists and medical doctors; the women were not recognized or encouraged to become research leaders. Rather, they were expected to undertake facilitative and pastoral roles but these duties were not highly regarded by their bosses.

Such support and relational roles are gendered, as Ford and Harding (2010) argue, and cultural expectations surrounding gendered roles make it more likely that women will undertake such tasks: 'Women, in conforming to the appropriate configurations of gender have a higher percentage of contacts with people . . . spend more time . . . on administration work; spend more time communicating; tour buildings and care for the physical environment more' (Ford and Harding, 2010, p. 505). In taking on this hidden work, some women may suffer exploitation, as Newman (1995) suggests, because they take on much more than is asked of them, including working long hours. Furthermore, these long working hours undertaking support activities are expected of women but represent little value in career terms, in contrast to the long working hours undertaken by male scientists in pursuit of science and their science careers.

Nina, in a post-doctoral position, describes one example of lack of acknowledgment of management responsibilities undertaken. She complained that she had taken on running the department, in contrast to a male colleague who was single-minded in his approach to his research and continued to write scientific papers, ignoring the management and support work required. As publication records are a significant way that scientists are assessed, this presented a significant advantage to Nina's male colleague. During maternity leave, many women are disadvantaged by publishing fewer papers and this became even more of an issue if they took extended maternity leave to raise a young family before coming back to scientific work (Greenfield, 2002a; see also Delamont, 2003; Hosek et al., 2005). Nina, as a mother of a young child and pregnant again, was experiencing an additional burden:

> I'm more likely to sort out something that needs sorting than to say, 'I'm gonna write my papers!', whereas the chap I share an office with is very happy to say, 'Sod that to everything!' and sit there and do nothing but write papers all year. And so, obviously, his publication record is fantastic. (Nina)

Of course, this male post-doc can only write his papers all year if another scientist is prepared to take on his share of managing the necessary support tasks. Nina's male colleague was allowed, even encouraged, to pursue writing papers and concentrating on his area of scientific work. No one apparently challenged him to maintain the accreditation records, undertake health and safety audits, order laboratory consumables and ensuring that the laboratory was kept tidy as Jackie, Ella, Nina and others were expected to take on. Nina continued by describing in some detail how she contributed to the organization of the laboratory but to the detriment of her own research:

If somebody has a problem with their machine they'll ask me rather than him. One of the first things I did when I joined the department was to instigate a lab tidy rota because I'd never worked in any laboratories which were so filthy and quite frankly dangerous. (Nina)

Arguably, Nina noticing the 'filthy' laboratory labels her a 'm[o]ther'. The historical legacy is strong for support work to be associated with women and Nina was carrying out the gendered role expected of her in society and in science. Such a legacy acts both to devalue support work and to reinforce the cultural assumption that it is women who are the 'natural' support workers, who, as in nursing, tend to take on the 'role of "hand-maiden" to male professionals' (Witz, 1992, p. 68).

It was not surprising, therefore, that many women were ambivalent about undertaking support work. Most of them even quite enjoyed it but they realized that doing support work didn't get them recognized as the scientists they wanted to be. Jackie, for instance, was similarly willing to take on a support role in the laboratory but knew that this work would not help her in her career as a scientist and might even disadvantage her:

I'm very organized and I've taken on a lot of other roles within the laboratory because we're going for accreditation at the moment and I've done a huge amount of work towards that [extra set of responsibilities listed], those are all the roles I do in my lab and I'm not trying to boast but I do take stuff on that I don't have to. I do feel undervalued and I don't think my achievements are seen. (Jackie)

Thus, Jackie's contribution to the smooth running of the laboratory was taken for granted but added little in terms of the recognition of her scientific advancement.

Even women recognized as being good scientists took on additional unwanted roles to support their male bosses. Ella's medical boss expected her to write his research grants, which he did not find satisfying to do himself, and whilst it could be considered that this was a form of coaching, Ella felt put upon:

I was doing what 'Philip' [Ella's medical boss] wanted me to do at that time – largely his grants and doing all his dirty work whilst [also] running the laboratory. (Ella)

Ella's story is told in more detail in Chapter 5, but in brief, writing Philip's grants may have been given to Ella as a development opportunity but was not presented as such. She also carried out additional responsibilities and was expected to continue to undertake management duties alongside her research work, and was even expected to supervise other researchers. She undertook many of these tasks in her own time so that she could continue with her research activities:

> At the same time as I ran the research department, I was also expected to manage the lab. I did stupid things like I went to (Midlands town) for three months [to learn about a new technique] so I used to work there during the week and then come back at the weekend and go into the lab and do all the ordering on a Sunday morning and sort out all the invoices and do silly things like that. (Ella)

As Eichler (1980) describes, Ella was submitting to the cultural pressures on her as a woman. She did not expect to challenge or negotiate with her boss about the additional work she was expected to do and he had not suggested arranging cover for her while she was away. To Ella, it was the price of doing research work and she accepted it, albeit somewhat unwillingly.

Will, a senior scientist in a public sector organization, voiced his expectation that women rather than men would undertake support work by reflecting that women were more suited to certain work than men were. In Will's view, women were more suited than men to microscope work because women had more 'patience' to sit 'for several hours':

> The type of work that we've done for years in this lab is more suited to the female than to other ... [Valerie's interjected at pause, 'In what way?']. Well patience sitting at a microscope for several hours. Over the years we've tended to employ women but these are sort of technician grades at various levels for that particular type of work. (Will)

Will used the word 'technician' in a somewhat derogatory manner to imply that women were employed to carry out low-grade tasks and had little ability for anything more demanding. In other words, his statements show how he worked within Valian's (2004) 'gender schemas' in science, where he would 'underrate' women's ability and in so doing would exclude them from positions of influence (p. 208; see also Miller, 1986), placing them at the lower levels in the science hierarchy.

In summary, the hidden work taken on by women in being relational, caring, serving and supportive to carry out basic necessities in the laboratory was associated with the role of being m[o]ther, a taken-for-granted 'othering' role that is appreciated as a necessary function but has little status. There was an expectation that women would take on these subsidiary roles and that men need not be bothered with the minutiae of laboratory life. Hidden work was not recognized as valuable scientific work in these masculinized scientific organizations as it did not contribute directly towards scientific outputs. Women healthcare scientists were thus disadvantaged in career terms and were kept in their subordinate place.

OPPORTUNITIES OR CONSTRAINTS?

Many of the women in our research did not have access to the range of critical work experiences required to reach senior positions in organizations, as identified by Morrison et al. (1992) with regard to management posts: being accepted, being supported, receiving training and the chance to develop by being given challenging work. Such work experiences are equally important in scientific careers but mostly there was a lack of access to them for the women interviewed and they appeared to be more available to the male scientists in our sample.

The interview texts demonstrate the difficulties women have in identifying and accessing opportunities in healthcare science and the women were rarely able to make free choices to escape the norms and sanctions imposed by bosses and organizations, as described by Maranda and Comeau (2000, p. 38) and Kerckhoff (1995). In line with Judi Marshall's (1995) research, several women were manoeuvred into taking options that suited their bosses but disadvantaged the women in their own career development where they were either moved away from science or from the area of science in which they were working. Women themselves appeared to rationalize the moves even though they had been sidelined from the research arena, the most influential aspect of their work activities (Ashcraft, 1999). The women were subtly eased out from their 'place' in science into a lesser 'place' on the sidelines of science such as administration, quality or project management.

One example was Margo who was in her late forties. Margo had considerable success in her early career but her fortunes deteriorated as her bosses later obliged her to change direction. Margo was born outside Europe but took her secondary education in the UK where she gained her first degree, including a one-year laboratory placement. Her boss gave her some sensible advice:

> He said, 'Have you ever considered doing a PhD?' I said, 'No' and he sat me down and said, 'Look there are two ladders here – people with a PhD come here and go up this ladder and if you don't have a PhD you're on this ladder and this is how you progress and it then takes a lot of effort to jump from this to this'. So I had never thought about it but he gave me the confidence I guess to say, you know, have a go and do it. (Margo)

This prudent advice encouraged Margo to study for her PhD in a post in a health-related industry where she could study for it as part of her everyday work. Margo considered that she had reached her current senior position because she had taken opportunities as they materialized:

I've made the most of every opportunity that's come my way. (Margo)

Margo progressed in her career, becoming internationally known for her work in a particular area. Her international career did not last, however, and Margo's bosses 'asked' her to become a scientific project manager for commercial projects, which as a consequence indirectly led to the deterioration of her international reputation as a research scientist. Her bosses apparently placed little value on Margo's scientific contribution and potential. The so-called opportunities offered to her by her bosses did not, in practice, advance her career in science. Rather, she was channelled into an administrative route and she was obliged to give up her international work, illustrating her lack of control of her career:

> I was an expert in my field and I used to attend a lot of overseas committee meetings representing the UK but over the years because I've moved away from my science subject and the overseas things and I have more skills now in project management and setting up new things and I think that's where they see my skills. (Margo)

Margo rationalized the direction of her career away from science when she said 'that's where *they* see my skills'. Her bosses were not apparently interested in her own development and her loss to them as a scientist was secondary to her worth to them as a project manager. In accord with Miller's (1986, p. 9) view that both dominant and subordinate groups favour avoiding conflict, Margo did not challenge her bosses. If Margo noticed the subtle masculinist behaviour and actions of her bosses, she took them for granted and, like most women in the study, did not try to change their minds. The changes crept up on her. Women are constrained from negotiating. They ask for less than men and get less; if women do ask for more, they are condemned for behaviour unbefitting a woman (Babcock and Laschever, 2003; Bowles et al., 2005).

Margo accepted the role of introducing and running new projects and she was successful at it. Although project management was not her first choice, she saw it as an opportunity:

> I would say that that's how it's been for me because opportunities have arisen and I've just sort of grasped them with both hands whether I was comfortable or not and just made the most of them and I've been very lucky. (Margo)

By offering this so-called opportunity to become a project manager however, Margo was channelled away from her scientific career. She became invisible as a scientist and she was kept in her 'place' by being side-tracked into managerial and administrative roles. Even though such roles conferred a degree of seniority, they led her away from her international role and concomitant opportunities to become a leading research scientist. Despite this she maintained a high work commitment but perhaps lost the high aspirations identified by Kanter (1977) as a characteristic of people who are able to access opportunities. Although Margo considered that she was valued by her employers for her administrative skills, she was not so highly valued as a creative scientist, which prevented her from progressing to the next level in her scientific career. As Maranda and Comeau (2000) point out, 'people who are concerned about job security can be manipulated by playing on their fear of losing their livelihood' (p. 43).

Margo described how she was asked to set up new projects that had an element of uncertainty about them in terms of their likely success, although she was somewhat self-effacing about her abilities:

> We won a contract and I guess they needed someone to set up the facility. I don't know to this day why I got pulled out of what I was doing and asked would I like to do this to, which I thought, well, this should be an interesting challenge. I knew nothing about [the subject] but I was willing to have a go. (Margo)

So Margo set up a risky science project to expand the business because she liked the 'challenge':

> I guess in a way that's me. I like the challenge of something new and doing something different and it was a success and then we expanded and I ended up in business development. (Margo)

However, in the process of setting up one new project, the risks materialized and the outcome was not as good as anticipated. Allocating Margo a 'risky' project is in keeping with observations by Ryan and Haslam (2007) that women are often appointed to senior roles only when these comprise elements of high risk (and are consequently perhaps not attractive to male candidates). The result of the poor outcomes of the project was that Margo lost her senior position:

> I set up the X labs. I'm not sure how I got involved with it as I'm not really a specialist in that field. I think it it's a case of what opportunities arose and what people thought I was capable of doing. I actually became head of X at one stage but it wasn't making any money so we decided to close it down and set up a different group. (Margo)

Margo was not asked to lead the new group and seemed to accept the decision as being something she was part of by saying '*we* decided', indicating that she thought she was part of the decision-making process

despite losing her leading scientific role. She had been used by her organization to its advantage by being asked to continue to set up new projects
but in so doing she lost her position as a scientist and was kept in the subordinate place away from the masculinist science place of action. Her previous scientific expertise counted for little. In keeping with Valian (2004,
p. 208), Margo was subject to the 'gender schemas' in her organization that
'skew our perceptions and evaluations of men and women, causing us to
overrate men and underrate women. Gender schemas affect our judgments
of people's competence, ability and worth'. Margo thus lowered her sights
in terms of advancement either as a scientist or as a business manager,
accepting her subordinate place:

> I think at one stage I would have said, Oh yes, I want to be on the EC [European
> Commission] committee and head of department or whatever but I don't think
> that's my ambition now. I think now I want to enjoy what I do rather than be
> ambitious and seek a goal up there. (Margo)

So Margo offered a reasonable interpretation of why she had become less
ambitious and was now unlikely to become a 'head of department': she
was going to 'enjoy what I do'. Our interpretation, however, is that she
had been passed over for promotion on account of her gender and was
prepared to rationalize her marginalization by saying, 'I enjoy what I do'.
The actions of her bosses were not overtly discriminatory but prevented
her progression as a scientist by subtle masculinist hegemony.

Furthermore, Margo had previously been regarded as an expert but by
moving in a direction away from science she lost the expert 'knowledge'
and became associated with the 'experience' as a project manager that
had less value and positioned her as an administrator (albeit a senior one)
rather than a research scientist (see Code, 1991 as discussed in Chapter 2;
Battersby, 1989). Her choices were limited. According to Archer (2000),
individuals consider they make 'active' choices but to do so people would
need to be able to predict their 'preferences', which recognizes that making
choices may not be successful (pp. 69–72). Margo was not able to anticipate her preferences and act on them and could not be termed an 'active'
chooser – she was manipulated.

Experiencing subtle masculinities in action, Margo was excluded from
a leadership position associated with men because she did not fit their
expectation of an expert or leader and her role as expert was removed from
her. Later communications with Margo indicated that she had been asked
to head the quality department in her organization, an important but less
prestigious area than leading-edge research and often populated by women
who move from science careers to ones on the periphery of science.

We now look at what happened to the women in our study who started work without a PhD who subsequently showed an aptitude for research. Healthcare scientists who do not have a PhD before they start working in healthcare science but later develop an ability and interest in R&D need to gain this level of qualification as an important next phase in becoming recognized as a research or clinical healthcare scientist. To accomplish this academic step, registration as an external student with an appropriate university is required to become a part-time student who is also employed. To achieve this stage, a supporting medical or a senior scientist boss sponsor/advocate is needed who will have contacts at universities and who will also advise on an appropriate research topic and supervisor. The process of achieving a PhD whilst being employed seemed much more problematic for the women in our study than for the men, and Carla, Mary, Rosa and Maggie were all unsuccessful in making this advancement whilst Tim and James seemed to find it easy. Jackie was initially unsuccessful but eventually took a different post where her bosses were more sympathetic to her ambition of achieving a PhD. When followed up five years later, none of the other women had obtained or were working towards a PhD and two of them had taken early retirement.

Having described Margo's experiences, we now consider Jackie's situation, a younger woman who was at the beginning of her career as a healthcare scientist employed in a public sector institute. During the initial stages of her interview, Jackie appeared to be supportive of her employer, indicating at first that she considered that her public sector organization offered equality of opportunity:

> I think the opportunities in [her organization] regarding women are very equal. I haven't personally come across non-equal things towards women. (Jackie)

However, as she relaxed, she later modified her viewpoint, saying:

> A lot of the higher roles are male dominated from what I can see and I do think there is a sort of male dominance where all the higher people are male. (Jackie)

Following gaining a First Class Honours degree, Jackie started working in an R&D laboratory and progressed well. She was offered the opportunity to study for a PhD in the distant future by her boss but this seemingly attractive offer was not genuine and was later subtly withdrawn. At this point, in her mid-twenties, it seemed that Jackie had taken on too many additional responsibilities, including the hidden work of preparing the laboratory for accreditation; Jackie was valuable to her boss in this quality role but it did not mark her out as a scientist. Instead, Jackie's boss initially offered her the opportunity to study for a Master's degree in a quality-related subject with the promise that she could then progress to do a PhD provided all went well.

For Jackie, although she worked in R&D and wanted to do a PhD, she was then advised by her boss 'not to do a PhD at that stage but to look into the route of registration with the HCPC' (see HCPC website), which would become linked to her becoming registered as a biomedical scientist with a protected professional title:

> I was advised not to do a PhD but to look into the route of registration – the main thing was that there are more opportunities with registration because it's internationally recognized. (Jackie)

This advice given to Jackie by her boss was inaccurate, as registration with the HCPC is only applicable within the UK, whereas a PhD would be a more transferable qualification internationally. The suggestion to become a registered biomedical scientist was to her boss's advantage but not to Jackie's as she wanted to be a research scientist but didn't know how to do it. Jackie's boss probably recognized that it would be easier for her to become a biomedical scientist as the routes for study and qualifications are in place through the Institute of Biomedical Science (IBMS) and he would not need to be involved to provide support. Once qualified, transferring from a biomedical scientist position to that of a research clinical scientist would mean transferring from one career ladder to another similar but parallel career ladder – the career ladders do not meet and the gap is very difficult to bridge. Having a PhD would be Jackie's passport to this other career ladder where she could be recognized as the research scientist she wanted to be. Furthermore, being registered as a biomedical scientist would not necessarily help Jackie if she moved away from her institute as she would find it difficult to gain a post in a clinical diagnostic laboratory due to her limited experience in processing clinical specimens in her current organization.

Although others in her organization had received funds to help pay for part-time PhDs such an offer was not suggested to Jackie, who was thus disadvantaged compared with male colleagues. Much as Jackie wanted a PhD, she could not afford to study for it full-time on a grant as it would mean she would be obliged to give up her secure post. She assessed the options available to her:

> So, you know, I thought about it carefully and looked at my options. I looked at where I could go with a PhD and where I could go with registration and I thought registration for me would be a very good thing; I didn't want to give up

my permanent position to do a PhD full-time because I couldn't afford to pay the rent. (Jackie)

Another pressure was put upon her. Even working for her HCPC registration proved difficult as her boss changed his mind about this as well as about studying for a PhD. A young man joined the department who was already registered with the HCPC and Jackie's boss suggested that the laboratory did not need two registered people:

> At the time there wasn't anyone in the department registered and then 'John' came who is registered so he [Jackie's boss] said maybe Jackie doesn't need to be registered after all. (Jackie)

Jackie fought for this option and eventually her persistence paid off and he relented, not about her studying for a PhD but about studying for HCPC registration. So Jackie went on to achieve her registration with the HCPC by part-time study over two years and became a biomedical scientist. However, despite having a First Class Honours degree and working in an R&D post, she was kept in a constructed isolated and subordinate place, labelled by her boss and some of her colleagues as 'other', different and a semi-professional (Etzioni, 1964, 1969; Hearn, 1982). As an outsider, Jackie was not invited to aspire to the ranks of research or clinical scientist and was a target of demarcationary professional barriers (Witz, 1992) from a male clinical scientist boss.

Jackie saw her ambition for a research career fading and the chances of developing herself diminishing. As a consequence of being in unworkable situation, Jackie was finding her post unbearable:

> I really do like my job and I would love to progress in the sort of work I'm doing but I don't feel like there's room there for me and I've actually been told on occasions that it's not my position to suggest things. They don't really want me to move up so I don't see how I can stay when I feel like I'm being held back but I don't know where to go. (Jackie)

As Fotaki's (2011) study describes, where a woman's comments are quashed, and similar to Angela discussed earlier in this chapter, Jackie's verbal contribution suggesting improvements in the wider laboratory was not welcomed and she felt pushed down. Far from encouraging her to develop a research career, Jackie's boss suggested she would be better staying as a biomedical scientist and concentrating on 'quality' (as Margo eventually did), which would enable her to continue her contribution to the laboratory's accreditation. So Jackie gained her Master's degree by part-time study in a quality-related subject. Continuing his subtle masculine hegemony, her boss then avoided discussing with her directly her possible career progression as a research scientist, in line with Miller's (1986) observations that 'dominant groups prefer to avoid conflict' (p. 9). He also made provocative comments at meetings, suggesting that she wouldn't be able to find the time needed to study, despite knowing that Jackie's study time would be outside normal working hours. Jackie used the term 'fight' to describe her feelings – the word 'fight' was used five times in her interview – but Jackie never won a fight with this boss:

> [H]e mentioned, Oh, Jackie might not be able to dedicate time to [do] the extra work that she needs to do, so I do feel I've had to fight. (Jackie)

During the interview, Jackie was uncomfortable, nervous even, of admitting her difficulties and asked, 'How confidential is this?' in case she could be seen to be complaining, which she worried might get back to her boss. Jackie was oppressed by her boss's 'harmful' and 'gendered expectations' of her, where she felt undervalued and was excluded from advancement (Rhoton, 2011, p. 697), but she put up with them, having little choice unless she were to resign from her post (which she eventually did). Despite fighting for opportunities to study for further qualifications, as time went on, it became clear to Jackie that her career was not under her own control and she was unsure what to do next:

> It's difficult to lay out your career path or lay out where you want be. I mean you can always say where you want to be in ten years time but it's not necessarily going to be a route that you're able to take so now I find it difficult to think about what's going to be my next move. (Jackie)

Experiencing the subtle discrimination in her gendered workplace but not knowing the best way to deal with it, Jackie was considering the options open to her:

> I feel undervalued, this is what I'm saying when my negativity comes in because I feel deep down that management don't want me to do anything extra and they're looking at me as a trouble maker. It's not what's said but um that's how I see it underneath. I've had to fight. (Jackie)

Fighting for opportunities to study took a lot of time and energy but was hidden work and having a qualification in quality was not recognized as 'real' science and it did not lead to her advancement. The effort of fighting did not add to her scientific credibility, and she was overlooked as the scientist she wanted to be and so kept in her subordinate place. Jackie's career was blocked by her male scientist boss who removed access to opportunities, perhaps not wanting her to join him in the ranks of clinical scientists. Jackie eventually moved to another post in a different laboratory with a more supportive boss where she was able to study for a PhD (told in more detail in Chapter 5).

Carla, in her early forties, was another woman who had been unable, in her case, to complete a PhD she had started, and where access to that opportunity had been curtailed by a hegemonic male senior scientist boss. When Carla finished her Master of Science degree (MSc), 'David' (her boss) had at first encouraged her to study for a PhD and she had commented more than once in the early part of the interview how he had helped her with her career and had described him as 'absolutely dynamic'. However, as she relaxed in the interview she revealed that one year into her PhD studies when she had taken on additional management and other supporting roles to build up the laboratory, David changed his mind and required her to give up her work on her PhD and she had never started it again:

> David [Carla's boss] was great and absolutely dynamic. Although having said that he gave me the opportunity to do the MSc, he was the one that when I started a PhD for a year there was a bit of conflict about whether I could do the lab work [required] because I'd taken over so much management and he just said, 'Oh well you'll just have to drop the PhD then' and I didn't think that was very supportive after all. (Carla)

Carla found that taking on a more management role in supporting her boss meant she had to reduce the science element in her job. Carla was undertaking an important laboratory management role but she was not identified as a 'laboratory manager' by him and the leadership place remained reserved for men in a similar manner to the way Ella had been treated, as discussed earlier in this chapter. Her role was in supporting him in management tasks and, rather like Jackie taking on the accreditation tasks, it was more important to him that she took on these lesser roles than her being a scientist – so much so that he compelled her to give up her PhD studies. Although he had helped her with her Master's, he was powerful enough to provide barriers later when it suited him. This male senior scientist who made the offer could both give and take it away, illustrating the fragility of a paternalistic relationship for a woman recipient. She was identified as a support worker and kept in her place. Like Jackie's boss (and Mary's, as discussed in Chapter 6), it seemed that Carla's boss also resisted her joining him in the place of experts and leaders – he had built invisible and impenetrable barriers of professional demarcation by exerting his subtle hegemonic power that kept her in her place as a biomedical scientist as well as a woman, and she experienced the double bind described by Tannen (2008).

DENYING OPPORTUNITIES

Rosa and Maggie were similar to each other in that they each made significant clinical contributions in their different clinical diagnostic laboratories
but without having a clinical qualification. Both women were around 50
years old and had started work in the laboratories as graduates and each
studied for an MSc part-time. Rosa had also worked in R&D laboratories
and both had progressed through the grades of biomedical science and had
managed large laboratories, though without management qualifications.
Both were respected biomedical scientists registered with the HCPC, rep-
resenting their laboratories on national committees in their different areas
of expertise. Both were given limited training in order to sign out clinical
reports. Rosa commented:

> I suppose what I got from her [medical boss] was experience in terms of clini-
> cal scenarios but I suppose I also got just a passion for helping patients and
> [understanding] how important it was to get everything absolutely right. (Rosa)

Similarly, Maggie said she had always been more interested in clinical areas
than 'science':

> I've always been fascinated in the clinical side of things rather than the scientific
> side. It's the result at the end of it and how that affects the patient that interests
> me and I'm quite good at it. (Maggie)

Neither woman was encouraged or supported to go the stage further to
take the exams of the Royal College of Pathologists (RC Path), which
would enable them to be recognized for their clinical decision making.
Rosa commented:

> Everyone was very supportive. Once I said I wanted to do more clinical stuff,
> all our clinical consultants basically agreed for me to sit and do all the tutorials
> along with the [medical] registrars and I go to all their clinical meetings and I
> still carry on doing that. (Rosa)

Currently, apart from the medical route, doing a PhD first and then
achieving RC Path exams (primarily for medically qualified doctors
working in clinical diagnostic laboratories) is the only way a healthcare
scientist, whether a biomedical or a clinical scientist, can formally be
recognized to give clinical advice on laboratory test results in the UK. In
clinical diagnostic pathology laboratories, such a qualification is regarded
as being 'equivalent' to a medical doctor by the professions and accreditation bodies for certain aspects of the clinical diagnostic role and enables

non-medically qualified healthcare scientists to advise clinicians on treatment of certain infections.

Supporting the women to undertake further clinical qualifications and training would have taken considerable time and energy by the medical staff, which they were not prepared to undertake: the medical bosses separated themselves from those in the semi-professions by erecting invisible but impenetrable professional barriers of demarcation (Etzioni, 1964, 1969; Hearn, 1982; Witz, 1992). They effectively left these two women in limbo in career terms, in the place of women with no recognition for the responsibilities they were carrying out.

As biomedical scientists, neither woman was supported in her career by her medical bosses in the way that a trainee medical doctor or clinical scientist would have been in similar positions. Because they were biomedical scientists as well as being women, they manifested Laws (1975) 'double [or triple] deviance' and the 'deficit' of not being a male, a medic or a clinical scientist (Miller, 1986, p. xiii; see also Rosser, 2004; Wajcman, 1998). There is no pressure on medical bosses to develop the careers of biomedical scientists; that responsibility falls to laboratory managers, which is the natural career goal for biomedical scientists. Like a medical boss, a laboratory manager may also be resistant to a biomedical scientist pursuing opportunities for advancement in a different, more clinical and scientific direction. A qualification such as RC Path could eventually put the ex-biomedical scientist above the laboratory manager in the hierarchy because of the acquired clinical expertise, which is rated more highly than management or laboratory/science expertise or experience. In accordance with Code (1991), who notes that women's experience is valued less highly than 'masculine' expert knowledge, Rosa and Maggie were not encouraged to develop in the additional areas required to facilitate a move into a clinical career, as Rosa recalled:

> There weren't any openings for a junior clinical scientist and once 'Janet' [previous medical boss] had left the laboratory the opportunity didn't come up for me to do a PhD or I might have been tempted; I might have been persuaded to do it but I couldn't see who was going to be a reasonable supervisor for me. (Rosa)

This quote from Rosa suggests that she was badly advised. Despite spending several years working in R&D, Rosa was not encouraged by her medical bosses to do a PhD. Rosa was on the wrong career 'ladder' for a research career and could not move from one ladder to another, although she was capable; for her a PhD could have been the start of a research career and the RC Path qualification could have been the marker to help her progress to a more clinical role.

Rosa did not expect career support from her medical bosses and did not ask directly or negotiate for such help from them, and such support was not offered. She explained it thus:

> I thought about doing RC Path [exams] but it would mean that I would have to go and spend a lot of time in [another laboratory] and I'm not entirely sure I'd be released to do it. (Rosa)

Maggie also believed it would not be possible to study for the RC Path exams and did not seek it:

> I've not gone in that direction, no I have thought about it but it isn't really possible. It all comes down to funding because there would be training implications. (Maggie)

These two women were grateful to be given a degree of clinical responsibility and asked for nothing in return. Instead, both Rosa and Maggie studied for the highest-level professional exams set by their institute and both passed with very high marks:

> I suppose the only thing that I might have done was management and pushed myself forward for a higher grader earlier. I can't see how I can get a more clinical role without the RC Path exams. (Rosa)

Access to taking the RC Path exam would have given both Rosa and Maggie significant benefits and would also have helped their laboratories. For Rosa, reviewing her unsuccessful attempts to access opportunities and move on in her career to a more senior clinical position, she saw the responsibility for her lack of advancement as her own, seeing herself as being 'deficient' because she did not 'fit in' (Miller, 1986, p. xiii; see also Rosser, 2004; Wajcman, 1998). Rosa ascribed the 'fault' as her own and did not see that it was the result of the gendered nature of her position and the inequitable system in which she worked, where, as Witz (1992) describes, her medical bosses erected barriers of demarcation and she was kept in her place as a biomedical scientist. Rosa found a lack of appropriate opportunities to move on in science open to her either to do a PhD or to study for the RC Path exams.

Maggie, like Rosa, was denied opportunities that would have been available to a medically qualified doctor or clinical scientist. Like Rosa, she undertook limited clinical work but without formal recognition. Because she was a biomedical scientist and because of her gender, opportunities for Maggie to progress (like other women described in this chapter) were blocked. Nevertheless, when Maggie was asked if she had met any barriers in her career, she said she could not think of any:

Really I haven't ever felt there have been any barriers. I felt the people that I've worked for have always been very considerate and when you asked for something if you put your case forward it has always been considered fairly and I have had some good outcomes so I've never felt that as a woman I haven't achieved what I could have done. (Maggie)

Despite not being given the opportunity to study for the more clinical RC Path examination, which would have helped her gain recognition for the clinical work she was undertaking, Maggie accepted her place in the hierarchy, apparently considering that she had been treated fairly. Like Rosa she was unsure what to do next. She was willing to stay where she was, but knowing she could have achieved more had she 'started right':

Probably if I was starting again I wouldn't be a biomedical scientist I would probably have been a doctor if I'd started right. I'd have been an infectious disease doctor or something so I could have had more impact on patients. (Maggie)

Although she did not have children, Maggie's said her reasons for not seeking more professional qualifications were to do with family rather than accessibility to the study:

I've tried not to let my career take over my life. I haven't got any kids but I love my home life and I've got two dogs and I love going out for walks and I love travelling and I just like to keep the two completely separate. (Maggie)

Both Rosa and Maggie were clever women but their abilities were not recognized within their science environments. Maggie's language indicates her lack of confidence and she rationalized her lack of advancement as being of her own choice, rather than commenting on the barriers she met and did not try to overcome. Unlike Rosa, Maggie did not see her lack of progress as her own deficiency. She did not expect to be able to get the training she would need to get the RC Path exam that would give her recognition for the role and responsibilities she was taking on behalf of her medical colleagues; she realized that more involvement on their part would entail effort by them and she felt they would not be willing to give this. She accepted her woman's place in the semi-profession she was in as a biomedical scientist and did not fight to change it. Recognizing that she was in a weak and subordinate position, she did not try to negotiate with her bosses. Despite saying she liked to spend time away from working, Maggie, like Rosa, continued to study and passed the highest-level professional exams available to her set by her professional body.

So both Rosa and Maggie were kept in their place by lack of opportunity and lack of a medical advocate to provide access to the required qualifications to develop a more clinically oriented career. Their ability was capitalized upon but their further career development was not encouraged. Studying medicine in her twenties may well have been an unachievable ambition for Maggie, but the professional barriers of demarcation and the subtle hegemonies prevented both Rosa and Maggie from advancing their careers in healthcare science even after they had proved their academic and professional ability in a suitable scientific environment. To recall the advice given to Margo by her first boss, it takes a lot of effort to move from one career ladder to another without a PhD, even in these two cases where the careers of biomedical and clinical scientists are closely related.

Rosa and Maggie's male medical consultants were content for the women to assist them with their clinical responsibilities but not to the extent of helping them develop their career by studying for a PhD or for the RC Path exams. The formal clinical training available to clinical scientists such as Carolyn was not accessible to Rosa and Maggie, both biomedical scientists, despite them both having similar clinical responsibilities. They were encouraged to take on these responsibilities but discouraged from progressing their careers formally in that direction by the subtle hegemony and barriers of professional demarcation (Etzioni, 1964, 1969; Hearn, 1982; Witz, 1992) of their medical bosses, who presented their decisions as benefitting the women rather than themselves.

Knowing her 'place', the perception of the barriers, constraints and notions of confinement were enough to stop Rosa thinking she could be 'released' to gain experience in another laboratory, which would be necessary for her clinical training. Rosa was typical in being unable to negotiate a change of direction and had not even asked directly, concerned that she would be going against the social and organizational norms expected of a woman and a biomedical scientist. In this way she provided a good example of how women lose out by not considering the possibility of negotiation and have no expectation of being offered the opportunity to negotiate (Babcock and Laschever, 2003; Bowles et al., 2005).

THE MEN

With regard to the men's experiences, only a few men seemed prepared to discuss benefactors or how they accessed opportunities. Frank, a medical doctor and CEO in a public sector organization, said: 'It's not been a planned career put it like that, sort of reacting when opportunities arise'. For him, there was a laidback approach to taking opportunities:

I take it one step at a time. Each time an opportunity arises I say is that something I would like to do. Oh yes, I'll give it a go, without worrying too much if I don't get it. If I take it and I don't like it, I'll move on and do something else. (Frank)

Ralph, a CEO from a healthcare scientist background who started work with a PhD in a research post, was similarly confident that if a new job he had taken didn't suit him then he could find something else:

I felt that if it all went belly up then I could find something to do that would be of interest. I had no ambition, I had no idea I'd want to be doing this in five years time. (Ralph)

Twenty years younger than Ralph, Ben, similarly started work with a PhD in a public sector institute. He was another good example of a man with a relaxed and confident attitude to taking opportunities and developing his career:

[E]ven though I had an interest in science I didn't really have a plan of my career, never really have done, still don't really. So when one of my lecturers offered me a PhD at the end of the exams, he mentioned he had some money and a PhD so was I interested? So I kind of stumbled into it. I said yeah why not, I would give it a go. (Ben)

So for Ben, opportunities had been offered to him and money had been made available to study full-time for a PhD rather than him having to fight, or even ask, for openings. As with nearly all the men interviewed, choice and opportunities were easy to access, in contrast to the situation of most of the women healthcare scientists in the study.

Tim and James were senior healthcare scientists, both directors of laboratories with clinical and research responsibilities. Both Tim and James left school at 18 and started working as biomedical scientists in diagnostic laboratories where the usual route for career progression would be laboratory management. Like several of the women interviewed, they studied part-time for professional exams with the IBMS. Both were particularly interested in their different scientific disciplines and received help from paternalistic benefactors. Unlike Jackie, Mary, Rosa and Maggie discussed above, both men were *offered* the opportunity to study part-time for a PhD in their diagnostic laboratories, achieving their PhDs before they were 30, because, as Tim commented:

They knew I'd leave if they didn't let me do a PhD. (Tim)

James commented:

> I worked bloody hard for the opportunity to do that PhD. I stayed late and
> worked on little projects that nobody else wanted to do. My boss at that time
> knew I'd deliver what he wanted and that I'd get on. (James)

They were both also offered support from their medical benefactors to
achieve the qualifications of the RC Path, which enabled them to take on a
more clinical role as clinical scientists.

Frank, Ralph and Ben were able to choose which opportunities they
would take, implying that these were plentiful and, if the first opportunity
did not match expectations, there was always the likelihood that something
else would become available. The two men, Tim and James, who started work
as biomedical scientists grasped the opportunities offered with alacrity. By
contrast, the precious opportunities offered to women in the study were few
and far between. Furthermore, several women including Margo had little
choice but to 'accept' so-called opportunities but that were not ideal because
there were no better alternatives on offer and they had little choice.

MAKING CONNECTIONS

In conclusion, both women and men articulated subtle forms of discrimination that are normalized within organizations. Men were generally
described by the women and by themselves as being in the controlling position and women in the subordinate positions, but all parties rarely queried
these hierarchies. Because of these taken-for-granted subtle masculinities,
Valian's (2004) gender schema was endorsed and ensured that men were
judged to be more appropriate for advancement to the top jobs in science
than the women. In accord with Valian (2004), men overrated and supported other men and simultaneously treated women as though they were
subordinate to them. Women appeared to be left with no alternative but to
bear their place in the hierarchy and rarely challenged the behaviour and
actions of their male bosses, even when the women didn't like what they
experienced. The ways they found of coping (such as Jackie being obliged
to present a 'submissive' front) reinforced their subordination.

Male bosses typically helped younger men advance up the career ladder,
reinforcing Kanter's notion of 'homosocial reproduction' (1977, p. 63)
where men were praised for their confidence rather than their scientific,
management or people skills. Conversely, women were identified as offering relational skills that were regarded as ineffective (Fletcher, 2001, p. 3).
Bosses did not appreciate them for managing staff in a caring way – rather,

their relational style was disregarded and was part of women's hidden work. Women were expected to undertake the support roles that were on the edges of science; they took on these roles because it was expected of them and mostly they didn't mind doing them but they did not add to the woman's scientific credibility. Even when women were praised for their scientific skills, it was because they worked well under direction.

Fighting for opportunities for education and training was also part of the women's hidden work. It took a lot of time and energy but was not recognized as 'real' work. Again, the struggle did not add to their scientific credibility, and they were overlooked as the scientists they wanted to be and so kept in a subordinate place. In a process similar to that described by Witz (1992), several women biomedical scientists were trapped by exclusionary professional barriers erected by the more powerful medical doctors and senior clinical healthcare scientists who blocked the progress of the women's careers and reduced their access to opportunities.

Several women were remarkable in that their bosses presented changes that they wanted as opportunities for the women but the so-called opportunities didn't actually benefit them. Rather, they disadvantaged the women and they were moved away from their science speciality into areas they didn't particularly want to work in. In Margo's case this meant losing her international position and becoming a project manager and subsequently a quality manager, assigning the direction of her career to her bosses. Furthermore, bosses sometimes offered the women beneficial opportunities such as the opportunity to study for a PhD but later changed their minds and withdrew their offers, which of course further disadvantaged the women.

Women were invisible as scientists and were excluded by men from decision-making groups such as senior management teams, which were male dominated. Women found it difficult getting their voices heard and when they did, their views were not held as highly as those of men. Nevertheless, both men and women seemed to accept these practices and take them for granted – they rarely questioned the unfair practices. Although some women challenged individual male bosses when fighting for training, the women rarely negotiated well and they did not challenge the dominant masculinist discourse even if they were not happy about the hegemonic practices in their organizations.

Although all the women worked in science organizations where there were few women at the top, they seemed, like the men, to accede to the masculinist discourse as the norm and took it for granted that this was the way the science world worked. Few women or men seemed to notice the inequalities at the organizational level, although some women tried to challenge the unfairness as individuals. An alternative discourse among women was lacking. Perhaps not surprisingly, the women were resigned to the dominant masculinist discourse, having limited opportunities for challenging the norms of the masculinist science world. Even so, men accepted and women lived with the same taken-for-granted assumptions that men control the science world and women know their 'place' in it (after Newman, 1995, p. 19) – they seemed not to notice. This also resonates with Rhoton's (2011) work, except that her female interviewees, 'distinguish[ed] themselves from other women or practices and traits commonly associated with women or femininity' (p. 701) whereas our women interviewees did not criticize other women for being overtly feminine or for other characteristics – they supported them, for instance for being good managers, as Alice did. This difference could be partly due to Rhoton's interviewees being senior women ('full', 'assistant' or 'associate' professors, p. 700) in the United States and they may have needed to be more competitive in order to progress. Women in our study did not distance themselves from the behaviour of other women and did not belittle the feminine (cf. Rhoton, 2011, pp. 701–3). On the contrary, women were supportive of other women.

Although some women appeared initially to reject the idea that their disadvantages at work were due to their gender, their approach changed as they began to talk more, and became more open about their difficult experiences of subtle masculinities in action. Significantly, however, they did not allude to feminism being part of their thinking at all. Perhaps one of the problems for women who become scientists is the lack of a feminist debate in most university science education. Most women educated in the sciences are unlikely to have had any exposure to feminist thinking, unlike those studying the 'arts'. The very absence of an alternative feminist discourse where women's voices can be heard perpetuates the separateness of women and reinforces their lack of power and influence. Perhaps women inherently understand from years of marginalization that men are not interested women's position and do not listen to women's voices.

In a similar manner to science, Miller et al. (2002) record how 'management' is seen as being 'ordered and rational' (p. 26) and note the lack of consideration of gender within curricula for Master in Business Administration (MBA) courses. After much discussion on how this could be rectified in one particular MBA course the authors cite how a session on men and masculinity was included. Although the authors comment that this example should not be taken as a 'prescriptive model of practice' (p. 28), some success was achieved as students became unexpectedly engaged with the subject of gender.

The men interviewed in our study manifested a confidence in their careers that was absent from most of the women scientists. For the male healthcare scientists, their confidence seemed to be associated with being men in the man's world of science with the expectation of status as an expert scientist and leader and because opportunities were offered to them. In nearly all cases, women were less confident because options were few and far between. It seemed that men could select what they wanted to do from attractive opportunities offered (which they accepted as if they expected them to be available) and they had a degree of autonomy (Maranda and Comeau, 2000, p. 38) where choices were plentiful and options easily accessible and barriers few or non-existent. They could take advantage of opportunities offered and were confident that if they rejected an offer something else just as good, if not better, would appear later.

By accepting the construction of the masculinist discourse, men were able to continue their dominance and the women did not have the opportunity to challenge men's authority, living with the status quo that inherent masculine power keeps women in their place and stops them from progressing as Kanter (1977) and Miller (1986) suggest.

NOTES

1. Some of the ideas on 'subtle masculinities at work' have also been discussed in a different form in Bevan and Learmonth (2013).
2. Subtle discrimination in various areas of science is, of course, a worldwide issue, even in countries where legislation should protect women from more blatant forms of discrimination – in the UK, the legislation is in the form of the Equality Act (2010). This is confirmed by various studies, for example in the United States by Kemelgor and Etzkowitz (2001), Valian (2000), and Roth and Sonnert (2011) who report that 'informal structures and anti-bureaucratic practices disadvantage both female scientists and non-scientists' (Roth and Sonnert, 2011, p. 396). Similar examples are also described in France (De Cheveigné, 2009), Finland (Husu, 2001), Sweden (Peterson, 2010) and Turkey (Küskü et al., 2007).

4. Secret careers

A resident man creates extra chores, more washing, higher standards for cooking, more organization to suit his schedule and women have to set time aside to be with him, to be attentive and accessible.

(Chandler, 1991, pp. 121–2)

SUBTLE MASCULINITIES AT HOME

This chapter explores the lives of the women scientists at home and how their husbands and families influence their lives (all the cohabiting heterosexual women were married to their male partners in our study). Lives of employed women are complex, as many are juggling careers, supporting a partner and caring for children. Consistent with Miller's (1986) arguments that women are part of a subordinate 'wife' group who are 'highly attuned to the dominants, [and] able to predict their reactions of pleasure and displeasure', the women in our study indicated that they were more conscious of humouring and supporting their husbands in their careers rather than expecting support from them to help progress their own ambitions (p. 10). Moreover, as Miller (1986) suggests, both the dominants and the subordinates in the relationships we report on needed to have a shared understanding of each other's relative positions to avoid conflict and sustain their relationships.

The issues women face when they try to balance a career with running a home and family are multifaceted. Most women took on the 'second shift' described by Hochschild (2003) where they took more than their fair share of responsibility for the organization of the home. In contrast to Hakim's (1995, 2010) views, we take note that the women interviewed were committed to their scientific careers whether or not they had husbands or children. Their commitment to their careers was closely linked with their social identity and how they valued themselves. In accordance with Gatrell (2005, 2006a) even women who worked part-time did not lose their ambition, showed commitment and worked long hours (see also Brannen and Moss, 1991; Potuchek, 1997).

We show here that not only do husbands exert subtle pressure on their wives in the home, but the constraints from husbands influence women's working lives and indicate some of the reasons why women have limited opportunities for prioritizing careers. This leads to the disquieting conclusion that the similarities between the subtle masculinities experienced in the workplace may be re-enacted and reproduced within some home settings where women's careers are not considered worthy of much attention, and the concept of women as 'other' is continued. When women work in the masculinized place of science, we recognize that women are regarded as 'other', and sometimes what may be termed 'm[o]ther' (which is explored further in Chapter 6; see Fotaki, 2011 and Cooper, 1992 for a discussion of the othering in the work context, drawing on Lacanian psychoanalysis [Fotaki and Harding, 2012]), by male medics and scientists. This 'otherness' is manifested by the perception that women lack being men, as Davies and Thomas (2002) argue: '[C]onceptions of being positioned as "the other", of not fitting in to the "masculinist ideal", were very prominent. For many women, showing feminine attributes is something that always needs to be managed, as it can be seen to silence their authority and their involvement in decision making' (p. 479).

We argue that heterosexual and married/cohabiting women with children are considered 'other' or 'm[o]ther' by their husbands/partners in relation to their careers, their difference identifying many women with the private place of home rather than as professionals, rendering their career aspirations invisible. We also argue that women in the home 'know their place' as they do in their paid work (Miller, 1986; Newman, 1995; Puwar, 2004). As in their career pathways at work, we see that women's choices may be constrained in their marital homes by their gender and also by structural barriers in relation to their careers, akin to those described by Maranda and Comeau (2000) and by Kerckhoff (1995). In particular, we observe how married women with children appear to maintain a low profile about their careers at home, to the point where they could be said, effectively, to conduct their careers in secret from their husbands. In two cases where women raised the profile of their own careers after they were married, the relationships broke down and they divorced.

So, perhaps more disturbing, it is apparent that a woman's commitment to her career, particularly when she has children, is overshadowed by her own, her husband's and society's construction that women's expected place is to take prime responsibility in the home. Furthermore, subtle sexism at home customarily depicts a woman's career as secondary to that of her husband's neediness and we can draw on Hacking's portrayal of 'unmasking' (Hacking, 1998, p. 58) to make visible the practice that husbands/partners also discount women's careers: *things are not what they seem* (Hacking, 1998, p. 49 emphasis in original).

This chapter proceeds as follows. We first look at how women scientists experience their careers in the context of their home lives with their husbands. We then see how they share (or not) the responsibilities for the home and family and how this affects their careers. This is followed by an exploration of the balance of power in the home and we subsequently draw some conclusions.

NEGOTIATING OVER HER CAREER WITH A HUSBAND OR PARTNER

In this section we delve into the difficulties women experience where the lives and careers of their husbands take priority over their own careers. Exploring how women deal with patriarchal husbands, we extend Babcock and Laschever's (2003) concept (who consider negotiations at work over pay or promotion) and investigate women's success (or otherwise) in the gendered negotiations in the home.

Not only do dominant and subordinate groups need to have a shared understanding of the relative positions of the other to avoid conflict (Miller, 1986) but Gershuny et al. (1994) add that couples negotiate over time a better understanding of each other's position so that each benefits by accommodating these. Thus, during the period before couples have children, the power balance between women and their spouses is likely to have arrived at an acceptable plateau. When children arrive, a new dynamic is reached that has the potential to become strained as women and men negotiate the effect of new arrival(s) into the relationship. In the 'traditional' family of the first half of the twentieth century, men were the wage earners and women stayed at home and had responsibility for managing the home and bringing up children. This balance is closely linked to Talcott Parsons' ideology where notions of masculinity in the home are related to heterosexual men being in paid work with salaries higher than wives (who, if they were middle class, often left the labour market following marriage in any case; see Gatrell and Cooper, 2007; Parsons, 1971). Paid work was, and remains associated with masculinity (Potuchek, 1997) and Maushart (2003) argues that men still fear loss of manhood if they are not in paid employment.

The women scientists in our study fall into the category where they were highly committed career scientists. However, several women scientists both at senior and more junior level appeared to 'talk up' the careers, earning power and status of their husbands' jobs, attaching importance to the association of their husbands' higher salaries with notions of masculinity and male self-esteem. Of the married women, none presented their careers as taking precedence over the occupations of their husbands. Even where women earned more than their husbands, they nevertheless underlined the importance of husbands' jobs. Such an approach is in keeping with obser-vations by Hochschild (2003), Maushart (2003) and Potuchek (1997) that in heterosexual households, even in the cases where women's salaries are higher, importance is still accorded to husbands' earnings in order to avoid upsetting the balance in relationships where men regard breadwinning as integral to self-esteem. Maushart (2003) uses the term for such behaviours as 'pseudomutuality' (p. 23, citing Bittman and Pixley, 1997). In contrast to Jervey (2005), who found that role reversal led to resentment by husbands, it did not seem that earning more than their husbands/partners created dissent within the marriages of these women in our study – but perhaps this was because the women were careful about prioritizing their partners' careers.

Women scientists in our sample worked hard to preserve the situation where their husbands' careers appeared to be dominant and, in order to explore this more deeply, we now look closely at how two of these women (Rosa and Essie) conducted their careers in secret from their husbands and how two women (Paula and Tina) coped when their careers challenged relationships to the point of breakdown.

Rosa, a committed healthcare scientist in her late forties with two teenage children, was having difficulty seeing what career options were now open to her. When illuminating why her husband's career took prece-dence over hers, Rosa cited the fact that her husband earned more than she did as one of the reasons why she was unable to progress her own career – because she earned less, she was positioned as the main carer who needed to look after him and the family. Rosa said that her husband's career came first because of his greater earning power and she presented the decision as being jointly made:

> My husband's got a much greater potential to earn more money than me, so even if I got to the top of my tree he'd still be earning more money. I mean that's one decision we have made as a family unit. (Rosa)

Maushart (2003) suggests that women 'marry[ing] up' for security reasons, is 'one of the most serious obstacles to achieving genuine equality within marriage today' (p. 77). Higher earnings mean that men are able to exert more influence in a relationship (Brannen and Moss, 1991; Hochschild, 2003; Vogler, 1998; Vogler and Pahl, 1994) and women's financial contribu-tion to the household purse is often seen as of lesser importance than the man's and allocated to specific household expenditure such as paying for childcare, which then becomes the responsibility of the mother – 'female work' as Hochschild (2003) describes it (see also Potuchek, 1997). As we observe above, however, appearances may be deceptive and cases where

it might at first seem that men are providing 'security' might not in practice reflect women's financial contribution within marriage. For example,
Hamilton (2006), in the context of small businesses, notes that women are
not recognized for their contributions to wealth: '[A]lthough women might
be playing an active role in the creation of wealth they often do not receive
the recognition deserved and may be marginalized in the management and
ownership of wealth' (Hamilton, 2006, p. 260).

Hochschild (2003) indicates that one effect of unequal pay is to force
women's careers into the background (or the career may cease completely).
Where women with children, like Rosa, earned less than their partners, the
discrepancy in income was a significant issue for the woman who wanted
to progress her own career. Each of these women invoked her husband's
greater earnings to rationalize why her career took second place. The
women who had taken a career break would also be unlikely to 'catch up'
with the earnings of her husband.

These observations are consistent with Charles (1993), who observes
that women lose out financially by having children and taking on an
increased share of household duties. It was also apparent that having
children meant that a woman who earned less than her own potential,
and less than her husband, was placed in a financially vulnerable position,
which fostered an environment where patriarchal practices could flourish.
Because such a woman was financially vulnerable, she might then find
herself in a weak negotiating position with her husband over developing
her career if the gendered process of negotiation labelled her as asking too
much, as Babcock and Laschever (2003) describe in relation to negotiations with employers over pay. It meant that women who tried to negotiate
with husbands over their own needs for a career, felt the (possibly unconscious) need to humour the men about the importance of their careers
because they recognized the importance of male earnings. Rosa was
effectively kept in a subordinate place because she relied on her husband
to maintain their high standard of living. This was subtle masculinities in
action in the home. Rosa's husband took for granted that this was a decision 'jointly made' and apparently disregarded his wife's need for a career.
This was borne out by the men in the study, none of whom 'talked up' the
career of this wife or her earnings.

In Arlie Russell Hochschild's book, *The Second Shift* (2003) she gives
examples where women either modify their careers or sometimes give them
up for the sake of their husband and family. One of Hochschild's interviewees, 'Ann', deprecates her own career achievements, despite being a
successful 'supermom' with a high-level career that she eventually gives up
(p. 25). It seemed that Rosa was like Ann in many respects. Rosa juggled
a responsible job and wanted to develop a high-level career as a scientist,

but this was countered by the demands of children who needed ferrying to and from out-of-school activities, as well as a demanding husband who accepted without question the support that Rosa gave him, their children and the home. Rosa seemed to conduct her career in secret from her husband and he was apparently content to discount her career in favour of his own.

Perhaps some women live their careers vicariously through their children and like Rosa would not consider moving because of the 'good schools' her children attended. Whilst this is not to suggest that children should be moved from school to school to satisfy the career ambitions of their parents, it was noticeable that several women would not consider a move for themselves but did not expect the same sacrifices to be made by their husbands. All these factors contributed to a complex life, where Rosa fitted in her career to the needs of her husband and family so that they were not disturbed by it.

Like Hochschild's Ann who failed to 'honour her potential', Rosa's career sat in the shadow of her husband's (Hochschild, 2003, p. 112). She did not want to move her children away from their good schools and facilitated her husband's daily commute of long distances by being the one who ferried the children to and from their schools and out-of-school activities. Rosa did not expect her commuting husband to contribute to this responsibility and, on the contrary, she rationalized that he was too busy to do his share. She seemed to feel indebted to her husband because he earned more than she did and had 'a more stressful job':

> Now I'm earning what he was earning 20 years ago or something. I know that he has potential to earn more money so logically it makes sense for his career to come first. He's also doing a more stressful job which has got more responsibility than [pause]. (Rosa)

Having made this statement, Rosa paused and then commented as an afterthought that she had not taken into account the work that she brought home to do at night, part of the hidden work that women undertake whereas men tend to be more open about the extra work they carry out (Glucksmann, 2005):

> Then I have to cook the tea and afterwards I start working if I need to do some work. (Rosa)

Not only did Rosa perform a 'second shift' as described by Hochschild (2003) where women are estimated to work an additional month per year on the many activities in the home that go unrecognized and that are in addition to their paid jobs, but she had a third shift of doing her paid work at home out of hours:

> I don't think he knows that I chase round taking the kids here there and every-where. I don't think he knows I do all of that. I don't think he has much concept of what time I actually get home on an evening and it's not very much more before him and then sometimes I have to go back to it. (Rosa)

In Rosa's case (as with others), it appeared superficially that she facilitated her husband in being the 'traditional breadwinner' (Potuchek, 1997, pp. 72–3) but there was a degree of hidden conflict about Rosa's role in her marriage that became increasingly apparent during the interview (Hochschild, 2003). Rosa was struggling to fit the role of 'employed home-maker' with her desire for a challenging career (Potuchek, 1997, p. 42), which she kept quiet about at home so as not to disturb her husband and family. In doing so she discounted the additional effort she put in at home when she worked after the children and her husband had been fed and they could relax, but she couldn't – she picked up the work she had started earlier in her workplace.

Rosa could be termed a 'committed worker' (Potuchek, 1997, p. 42) on the basis that although her husband's salary greatly exceeded her own, her own salary was over twice the national median for full-time employees (women and men combined, according to the Office for National Statistics [ONS], 2016). Her husband's career took precedence and her own career receded into invisibility, at least in the context of the marital relationship.

As Potuchek (1997) suggests, most men and women consider that men must be seen to be earning to support a family, even where both partners are in paid work (even if women's earnings are equal to, or greater than men's). It seemed with Rosa, as in the case of several other women, that there was an unacknowledged tolerance of being in a subordinate relationship where she did not negotiate for better terms at home (or at work). The 'lagged adaptation' described by Gershuny et al. (1994; see also Potuchek, 1997), where couples come to terms with each other's position, did not work in Rosa's favour but benefitted her husband. Any conflict present was hidden or resisted and Rosa remained in a position where her husband, children and home squeezed her career into the background. No one in the family seemed aware of this as an issue and certainly no one spoke of it – the hidden issues remained dormant. Even though her career was important to her, Rosa did not try to raise its profile to one of importance with her husband or her children – just as she had not with her bosses, as discussed in Chapter 3.

Arguably, Rosa knew her 'place' and underplayed her commitment to paid work, almost conducting her career in secret from her husband rather than take actions that might challenge the stability of her marriage. Chandler (1991) argues that some women (and men) prefer not to challenge those parts of a relationship that are unsatisfactory in case it threatens the very foundations of the marriage (Charles and Kerr, 1988; Hochschild, 2003; Maushart, 2003; Miller, 1986). While it was not evident that Rosa regarded her relationship with her husband as unsatisfactory it seemed that she was cognizant of her 'place' and did not wish to jeopardize the unwritten guidelines between them (Edgell, 1980; Mansfield and Collard, 1988). Furthermore, as Miller describes (discussed above), Rosa apparently accepted her position as the subordinate member of the couple in order to achieve shared understanding and avoid conflict.

Essie's situation was different from Rosa's in that her husband did not have a highly paid career. Except for maternity leave for her four sons (in their late teens and early twenties), she had always worked. She was initially reserved about talking about her career and reticent about saying that her achievements were in any way something to be proud of where she had moved away from science and over time had gained a senior corporate management post. She had worked with several different male and female bosses but felt that men held the power:

> You can only hope that men will change but why should they change? If you are number one, why are you going to give up power? There's no incentive for men to do that and in many ways women don't help either. Both men and women do not want to change things. (Essie)

Essie didn't want to 'blame' men nor did she have an expectation that men would take any responsibility to change themselves. Indeed, she appears to 'blame' women for 'play[ing] into the hands' of men who retain their dominant position. Essie highlighted another significant issue: men may fear the loss of their own authority if they consider the position of women. Why would men want to change if they would end up losing their authority? (See Hartsock, 1983 and Sismondo, 1995 for discussion of this argument in a Marxist context.) Essie recognized that men's dominance reinforces women's weakness:

> I don't think we will ever change this view that men are number one and women are number two. Women can absolutely play into hands of those men who hold the view that men are superior and so we can't blame men, it's ingrained in society. (Essie)

Essie accepted that men's authority is 'ingrained in society' and so in keeping with our argument that the dominant masculinist discourse renders women's needs invisible, perhaps she considered herself as part

of the 'other' that separates women from men. Furthermore, the power
balance active in Essie's home also involved her four, near-adult sons who
played a significant, albeit indirect, role in her career:

> I'm the mother of four boys you know. I've always had a career and they should
> see me as equal to their father but there's a subtle ingrained thing in males that
> makes them think they're superior to us and it just seems to me that in my own
> home life there is no way that my boys should think of me just as a housewife,
> but they do even though I have always had a career. (Essie)

So how had the situation arisen where the five men in Essie's her life concluded that they were 'superior' to her and to other women? In the above
piece of text, Essie draws attention to the 'subtle ingrained thing in males
that makes them think they're superior to us': she said she knew her sons
felt superior by the way her older sons behaved towards girlfriends and
the disregard displayed by her sons towards her career. It could also be, of
course, that she was overwhelmed by the numbers of men around her both
at work and at home and did not feel able to challenge their view. Again,
like Rosa, her opportunities for negotiating were limited and she seemed to
keep from her family the importance of her career to her.

Perhaps Essie did not want to be seen as displaying her achievements.
Essie's husband did not have a high-level career and it seems likely that
she compensated for his possible feelings of inferiority by downplaying
her own career and boosting his self-confidence. Thus, it seems that, like
the women with highly salaried husbands, Essie also had a secret career,
where she found it difficult to be openly proud of her achievements and
kept them hidden at home, perhaps to safeguard the masculinity of her
husband and sons.

It is therefore worth considering the part played by teenage children in
the balance of power at home. Older children could play a role in contributing to the house and garden work but do not feature in Hochschild's
(2003) research on the 'second shift'. Perhaps some women forgo their
careers because of older children as well as their younger children.

There is another possible reason why Essie led a secret career and to
this end it is worth considering how men who are brought up by women
with high-level careers may end up being sexist. Connell (2002) hints at a
possible influence in a report on Russian women. Novikova (discussed in
Connell, 2002) 'points out the important symbolic *place* [of women] in
Soviet Russian culture' (p. 25; emphasis added). He argues that Russian
mothers, especially mothers of sons, are 'symbolically identified' with
'sending forth sons-as-soldiers to liberate the world' (ibid.) and interestingly one of Essie's sons joined the army and one joined the police, certainly macho professions.

Both Rosa and Essie kept their own career achievements, aspirations and ambitions from their families, which may have been a subconscious way of not threatening the status quo and keeping the peace at home and preserving their marriages. We now go on to discuss what happened when Paula and Tina confronted their husbands about their drive for developing their careers and how their husbands could not cope with that challenge.

Wanting to pursue a career in science and lack of support by the husband to the woman's career was a significant factor in the breakdown of both Paula's and Tina's first marriages (Hochschild, 2003; Maushart, 2003). Paula's marriage broke down early on in her career (and marriage) because, she said, her husband could not cope with her keenness to study to develop a career. She felt she had married too young and that her husband was very immature; money was tight and pressures on the marriage were too great and they split up.

Paula, now a senior academic with a university chair, took maternity leave nearly 30 years previously for only a few weeks mainly for financial reasons. Paula's short first marriage left her with a baby. As the major earner in the household she returned to work when her baby son was only seven weeks old. Working in the National Health Service (NHS), salaries for biomedical scientists in pathology laboratories were low at that time and Paula needed to undertake on-call duties to bring in enough money to live on, relying on her mother, who moved from Scotland to the south of England to help her. Paula was very matter of fact, recalling the first few years of her first son's life, but said it was 'traumatic' at the time. She explained:

> The only reason I came to [city] was because I was married quite young much to my mum's disgust. She didn't want me to get married but I was quite headstrong and I came here with my husband who had a new job. We had major emotional problems and weren't getting on at all so I tended to put my energies into my work. (Paula)

As a way of helping forget about her marital problems, Paula used her love of science and her long hours of work to help her cope with the particular stresses when her marriage was breaking down. Her science career gave her something else to concentrate on other than her unhappy marriage, and as the major earner in her second marriage she went on to have three more children. Thirty years later within this long and stable marriage, Paula gained an academic chair, working full-time in a university.

Tina left school at 16 and worked as a dental nurse and then studied for her first degree through the Open University. She described how her marriage had broken down because her husband was unsympathetic when she

decided to change jobs to pursue a career in science, whereas he wanted a
stay-at-home wife:

> The biggest hold up that stopped me doing what I wanted to do was my then
> husband who wanted me under his thumb and didn't want me to go and start a
> career. He didn't want me to have a job at all, he wanted me to stay at home, so
> that's a big disadvantage being a female and having babies. He always regarded
> it as my responsibility if the children were ill; you know I'm the female, I have
> to look after them if they're ill so I felt like a second class citizen when I was
> married but that came from my husband, um, ex-husband. (Tina)

Tina rebelled and they divorced.

Hochschild (2003) records similar experiences and describes situations where rifts between the expectations of wives and husbands lead to
divorce. In Tina and Paula's situations there was no 'lagged adaptation'
(Gershuny et al., 1994). Charles and Kerr (1988) point out that patriarchal
practices continue because men expect women to deliver a service to them,
and women usually do that. Where women do not accept the patriarchal
practices but no change happens, the marriage breaks down, as with Tina
and Paula.

NEGOTIATING OVER THE DIVISION OF PRIVATE LABOUR

The relative importance of housework in the total work undertaken in
the family does not seem to have changed a great deal over the years since
Oakley (1974) pointed out that housework has largely been ignored in the
sociology of work. A number of authors have written about the difficulties that women face in balancing work and home lives, all of which record
that the responsibilities for looking after the house and family predominantly fall to women and are often a source of contention (Chandler, 1991;
Charles and Kerr, 1988; Delamont, 2001a; Gatrell, 2005, 2008; Hochschild,
2003; Maushart, 2003; Potuchek, 1997; Smith 1987; Williams, 2000).

Chandler (1991) records that 'in a pre-feminist era it was accepted that
women's power in public life was restricted, but women who controlled
the domestic decision-making were viewed as matriarchal' (p. 118) but she
notes that this view has become contested, and the organization of the
home is seen as having little value by society in general. Similarly, Smith
(1987) suggests that women's work in the home is not of their own choosing and they have little control over it: '[W]omen's work routines and the
organization of their daily lives do not conform to the "voluntaristic"
model or to the model upon which an agentic style of sociology might

be based. Women have little opportunity for the exercise of mastery and control' (p. 66).

As Ruddick (1980) argues, managing a house and children requires private power, personal strength and innovation. These qualities are frequently overlooked and women become powerless in a public setting and frequently devalue themselves. Ruddick (1989) challenges normalized ways of thinking about motherhood, suggesting that maternal work requires rationality and discipline but these are often ignored – 'the intellectual capacities she develops, the judgments she makes, the metaphysical attitudes she assumes, the values she affirms' (Ruddick, 1989, p. 24).

The assumption derived from the Potuchek's (1997) research is that women have a choice about whether they undertake paid work or not, whereas men have little choice but to work for pay, which would seem to lead to a further assumption that men will be more likely to develop a career, whereas this is not expected of women. Such an assumption was not borne out in our research where all the women and men working in science regarded their occupations as careers. The women with families, however, had the additional self-assumed responsibility, and expectations by their husbands and society, to be the 'homemaker'. Furthermore, many women take on additional home/housework during maternity leave and subsequently find it difficult to hand it back to their partners on their return to paid work (Gatrell, 2005; Maushart, 2003).

In heterosexual marriages, domestic routines are fitted around male schedules and career patterns (Chandler, 1991, p. 118). Men rarely took on the role of homemaker in our study. Only one man, Alec, identified with this role (Alec's story is told in Chapter 6) and one other man, the husband of Jane, a medical doctor, gave up one business and started a new business from home so that he could provide a constancy in their children's lives. The men interviewed seemed to enjoy being taken care of by the women in their lives (Haste, 1993; Miller, 1986), apparently unaware of the complexity of the lives of their wives, with little awareness of how the women managed to juggle their home and work lives.

We now look in more detail at how Elena, Rosa and Margo handled the private division of work in the home. Most women echoed Hochschild's (2003) term 'second shift' with regard to how much work at home they needed to do after their paid work was finished and also that husbands shared little of the home/housework, preferring the outside chores that they could do in their own time. Elena commented:

> Now that I'm back at work I find there's another full-time job waiting for me at home. (Elena)

However, she said she preferred to do the housework herself with input from her 20-month-old daughter, rather than so-called help from her husband:

> I tend to do the housework for quality purposes but my daughter is great with helping so we do this a lot. (Elena)

So, consistent with Hawkins and Dollahite (1997, cited in Allen and Hawkins, 1999), Elena's husband resisted meeting the more exacting standards set by her; she was left to do the chores herself with her tiny daughter who was more helpful to her with the housework than her husband, who rarely provided a useful contribution. Furthermore, Elena was already reinforcing the stereotype to her daughter of a woman being a homemaker and doing housework, and so the habits of society were subtly reinforced. Perhaps Elena also did not want to ask her husband for help in case it was seen as 'nagging', as Maushart (2003) comments about life in her first marriage (which lasted three years): 'Then it dawned on me how much I hated "asking", how much it always came out sounding more like "nagging". It would be years before the choice of the verb *help* would register as significant' (p. 26; emphasis in original).

Perhaps not unusually for a healthcare scientist, Elena used the terminology 'quality purposes', deploying, with irony, the management language of the laboratory to her home. She also illustrated two standards of 'quality': the one provided by her (and her tiny daughter) and the lower standard obtainable from her husband, which was not to her satisfaction. By doing so, Elena indicated an element of self-imposition, which was typical of the women interviewed where they looked after the house and family as well as having a career, as Crompton (1997), Hochschild (2003) and Maushart (2003) show. But perhaps the self-imposition was apparent only because the support from the husband wasn't there? Elena's husband was not offering help (and certainly not taking equal responsibility) and she did not demand it, possibly fearing conflict. Different standards are expected by society from women and men for the cleanliness and tidiness of the home. Men are not criticized in the same way that women are if their home is not up to the standards expected by society, so by perpetuating the housework as women's work, masculine superiority is subtly reinforced and taken for granted (Hochschild, 2003).

Most of the women in the study seemed in the habit of undertaking more domestic chores than their male partners (Delamont, 2001a; Hochschild, 2003; Maushart, 2003; Nutley at al., 2002). As Hochschild (2003) describes, several women used phrases such as Rosa's, saying he's 'helping me out' to describe the input from her husband around the house or with the children, Thus, putting the responsibility for the organization of the home on her own shoulders rather than her husband's (see also Delphy and Leonard, 1992):

> To be honest, I think it's really difficult. I mean my husband is always supportive and if I ask him to do anything in terms of helping me out or because I'm going away he will do it but at the end of the day most of it falls on me. (Rosa)

Thus, the input to housework was not equal and Rosa and other wives did not seem to expect it to be any different. As discussed above, Rosa's career was conducted in secret and she did not want to raise its profile, seemingly preferring to maintain a quiet life and undertake all the domestic chores to avoid conflict:

> We try and balance the family and work don't we? But there is the problem that it is difficult to balance everything. We probably don't help ourselves really. I think the only time I've ever seen it work any differently is if the partner, the male partner, allows the woman to go and follow her career and he maybe takes a backward step, which lets her go ahead. (Rosa)

Like several other women, here Rosa expressed some frustration regarding the manner in which the imbalance of home duties perpetuate and wondered how women might do more to change the status quo. However, Rosa also considered that she could only pursue her career openly if her husband facilitated it, in which case he would have to be willing in her view to take a background role but significantly he would also 'let her go ahead'. Although she was actively engaged in her career, like others the only way she felt she could pursue her career was in the background, in effect in secret from her husband. And all the while she took responsibility for the family and household responsibilities. Rosa took her children to and from sports practices in the evening, came home to cook the evening meal and then did more work afterwards to catch up on long-term projects for her laboratory:

> We've been going through this skills initiative recently and there just isn't enough time at work to do it. My husband doesn't seem to mind that I don't sit down in the living room with him after dinner. (Rosa)

It wasn't only looking after their children and the house that was undertaken more by the women than their husbands or partners (no one in the study had the additional responsibility of caring for aged parents). In the epigraph to this chapter, Chandler (1991) points out that women with husbands or partners are often involved in responding to the basic needs of their partners as well as those of the children: 'A resident man creates extra chores, more washing, higher standards for cooking, more organization to suit his schedule and women have to set time aside to be with him, to be attentive and accessible' (pp. 121–2). Few of the husbands seemed to take responsibility for the management of any of the housework and would 'help out' on their terms when asked. Rosa, like Elena and the other married female interviewees with children, undertook a substantial 'second shift' of taking responsibility for organizing children, husbands and housework (Hochschild, 2003). Rosa appeared to prioritize avoidance of conflict with her husband, rather than pressing to negotiate a better position for herself:

> Valerie: Does your husband ever come home early so that you can stay at work?
> Rosa: Yes, yes, like today he's helping me out or if I go away overnight he will
> . . .
> Valerie: Those are interesting words aren't they – 'he's helping me out'?
> Rosa: Alright. Yes. Well that's true, alright I'm asking him . . .
> Valerie: Does he take an equal share?
> Rosa: No, it's not equal . . .

In terms of negotiation, Rosa may have reached a common understanding with her husband – Gershuny et al.'s (1994) 'lagged adaptation' – but the terms for Rosa were less favourable for her than for her husband. Her own career did not figure very highly in his list of priorities and while she may not have liked this balance, she accommodated it perhaps in order to preserve her marriage.

Margo's situation was not unlike Rosa's with two teenage children, except that she lived in an extended family where her parents-in-law were able to look after the children when she had travelled overseas for her international work. Organizing the house and extended family though fell to Margo, and like Rosa she juggled family commitments. Margo appreciated her husband's 'help' with the household chores but like a lot of men, her husband did a lot less housework than she did, as Hochschild (2003), Maushart (2003) and Delamont (2001a) illustrate. Margo commented that her husband did the vacuuming when she asked him to and would probably do more but her husband chose to contribute in 'his own time'. Margo found it preferable to do most of the housework herself at what she viewed as a more appropriate time rather than wait till her husband felt like contributing: it was less stressful to undertake the tasks herself than to ask for support from others. Delays were not acceptable to Margo:

> He does help, he does, but when I go home there's always something to do and
> I've got to do this and I've got to do that and he's very good and I'm sure he'd

help more but it's just my frustration that he'll do it in his own time and I'm not sure I can wait but I try to leave the vacuuming to him. (Margo)

The input of Margo's husband to the vacuuming is consistent with the Hochschild's (2003) findings that men undertake work at home that can be done when they want to do it, and not the demanding and frequent routines. It is stressful for a woman who wants to avoid conflict to keep asking for input from their husbands that is slow in being forthcoming (Chandler, 1991; Hochschild, 2003; Maushart, 2003; Miller 1986; Potuchek, 1997). This enables the man to have 'more control over his time' as Hochschild (2003, p. 9) points out, so it is not surprising that women become frustrated – the likely consequence being that they will have less control over their own time. Like Rosa, Margo's frustration came from balancing her career and the expected role of homemaker together with inadequate input from a husband who contributed only what and when he wanted to. Nevertheless, in accord with findings by Delphy and Leonard (1992), women invariably refrained from openly criticizing partners when discussing their inadequate input into the household tasks, saying, 'he's very good' (see also Crompton, 1997; Gatrell, 2005, 2007; Gershuny, 2011).

The ONS records the time spent on household duties by women and men but there are no recent data, stating on its website that 'ONS has successfully carried out time use surveys (in 2000 and 2005) [in many countries] . . . ONS has not committed to spending any money on running a further time use survey'. In its time use survey in 2005, the ONS states:

While overall 92 per cent of women do some housework per day, compared with just over three quarters of men (77 per cent), the pattern is different for individual activities. Women spend more time than men cooking and washing up, cleaning and tidying, washing clothes and doing shopping. DIY repairs and gardening are however male-dominated. (ONS, 2006, p. 38)

And again using data from the ONS, Gershuny (2011) comments that: 'the fact that women still do more unpaid work and less paid, means that women still have less opportunity to accumulate paid-work-experience type human capital, and hence have lower expected wages, than otherwise similar men' (Gershuny, 2011, p. 12).

In accordance with Maushart (2003, p. 92), Margo's husband was one of several husbands who were happy to take over the 'technological tasks' that they associated with masculinity. The association with masculinity made the household tasks he undertook (vacuuming or gardening) more acceptable for him to do and in turn these were acceptable to Margo as she got some 'help' with minimal conflict. Almost as a rationalization of her 'frustration', Margo countered this with:

My husband has always been supportive of my career or whatever I wanted to do, so he has helped me. Without that I don't think I'd be doing what I've done. (Margo)

Margo undertook a significant burden in taking on the responsibilities of the 'second shift'. In agreement with Hochschild (2003) and Maushart (2003), Margo resented taking the brunt of organizing the home as well as her career and recognized the unfairness of balancing these two strands, whereas her husband only undertook a fraction of the home tasks when it suited him, and she harboured a 'grudge':

I think sometimes it's more important for females because they're trying to balance the career and the other life and I think if I have a grudge it's the fact that we have to do both, yeah. (Margo)

Margo's situation seemed to agree with Allen and Hawkins's (1999) statement: 'Some scholars argue that although mothers may be making most domestic decisions it may not be consistent with their personal wishes' (p. 203).

In summary and in agreement with the literature from Hochschild (2003), Maushart (2003) and Potuchek (1997), Elena, Rosa and Margo were examples of women who undertook more than their fair share of household duties. Men took on these tasks only on request and the standard of their work did not meet the higher expectations of their wives. The men's preference not to be as good as the women at household chores could be termed 'learned helplessness' (Seligman and Weiss, 1980) – they did not try to change the situation and benefitted by it. The men enjoyed the fact that their wives did most of the home/housework, taking it for granted, and women's role was constructed as homemaker.

WHO CONTROLS THE HOUSE AND FAMILY?

Why do women continue to take on more than their fair share of the home work of organizing the children and the house? Is it because, as Mansfield and Collard (1988) and Edgell (1980) suggest, the husband controls the household and delegates the organization of the house to his wife? Hetty Morrison (1878 [2010]) voices an early view: 'Accepting ourselves at the valuation of such men as these, that woman's place is in the kitchen, or, to word it more ambitiously, that "woman is the queen of the home," the right I ask for is that we be allowed to reign undisputed there' (p. 118).

Janeway (1971) takes this notion a step further by suggesting that women see the home as the place, and for many women the only place where women can exert power. Even in the twenty-first century is the home the one area where women can be in control, even if it results in more work for themselves? If not, why do wives take on the brunt of the work? Martin (1984) proposes another possible reason. She sees housework as, 'a cultural practice of great symbolic importance and not merely the performance of mundane utilities' (p. 25). In line with this, it seemed that the women in our study demanded high standards in their homes that their husbands did not contribute to in anything like an equal share. The women then, were perhaps prepared to take on the additional work because of its symbolic and cultural importance but, significantly, they also had little choice. As noted before, it is women and not men who are 'surveilled' (sometimes officially by health and social workers) regarding the state of their homes. In some cases, paid help was available but mostly women took on more than their share of the work in the home, whether it was housework, caring for and ferrying children about or satisfying the perceived requirements of their husbands for well-prepared meals and a comfortable life.

To take this aspect a stage further, we look in more detail at the interview texts of three women interviewees: Angela, Alice and Rosa. Angela earned significantly more than her husband. In their household, she described the division of labour as 'girls' jobs' and 'boys' jobs', recognizing gender stereotyping in action. Like Margo and Maggie's husbands, Angela's husband preferred to do the technological jobs, although he liked cooking as well:

> My husband likes cooking though he doesn't do a lot of the routine stuff and he also likes building things, which he does very well. I do the accounts – I am much better with money than he is. He handed over his pay packet to his mum and he was happy to hand it over to me so to speak: girls' jobs and boys' jobs. (Angela)

Angela described her 'control' of the household, but recognized that these might only be in the areas that her husband did not want to be involved:

> I manage the household and he is happy to let me do that. It is my decision about what we do and how we do it; it would be a nightmare to ignore that responsibility. I have to be in control. (Angela)

Consistent with Mansfield and Collard (1988) and Edgell (1980), perhaps Angela's husband was willing to let her think she was 'in control' but was in fact exerting patriarchal pressure over her; certainly he did not offer to take on the responsibility for organizing the home. Angela laughingly commented:

> Of course he might just let me think I am in control! (Angela)

Echoing Babcock and Laschever (2003), Angela implies here that negotiations with her husband had been successful because he humoured her
by letting her believe she was in control, whereas in fact he was delegat-
ing tasks that he did not want to undertake himself. It would seem that
Angela's husband had the better deal in absolving himself of responsibility
for organizing the household and managing the finances, for, as Miller
(1986) describes, 'delegation' is always likely to benefit the dominant group.

Unlike most of the women discussed in the home context, Alice, a
healthcare scientist, and Jane, a medical doctor (discussed in Chapter 6),
were the only women who said that their husbands took 'equal responsibility' with aspects of looking after the children. In Alice's case, this appeared
to be far from equal, although her husband did more towards the childcare
responsibilities than it seemed most of the other husbands did:

> I was probably number one arranging cleaners and gardeners so there were
> other people coming to do my job but I was arranging it, not him. (Alice)

Alice recognized her responsibility for the domestic arrangements by using
the words 'my job', commenting that it was her responsibility to arrange
the cleaners and gardeners (indicating in addition a combined level of
income above some of the other interviewees) or failing that, to undertake
them herself. Employing other people to help with the chores seemed to
be a way of Alice rationalizing her husband's reduced input at home compared with her own. In the areas of childcare where he volunteered, he was
an 'equal' partner. Those areas where he did not contribute to their care
(such as cooking) were ignored by Alice probably to avoid conflict, as a
number of authors describe (Chandler, 1991; Hochschild, 2003; Maushart,
2003; Potuchek, 1997). Alice's husband also had his own busy career, but
she was pleased with his input into family life, saying several times that he
was 'very good', despite his input being in areas that suited him and he did
not help 'much' with the shopping, cooking or cleaning:

> I do all the shopping and cooking but he is always caring for our daughters – I
> think that is unusual. I must say he's been very good at that. He was actually
> wanting to do it and he would also take other people's children, say to and from
> school or sports events. It was always him and various women doing that but I
> mean he doesn't do much cooking and not so much cleaning. I get the children's
> clothes ready and do the washing but otherwise the care of the children is about
> equal. He doesn't do much of the other things so I would say I have the primary
> role. I decide what we are going to eat and make sure there is food in the house
> and organize the shopping. (Alice)

Alice's praise of her husband seemed to be more because she wanted to
be pleased about her husband's input rather than the fact that he actually

took equal responsibility. Typically, as Gatrell (2006b, 2007) and Maushart (2003) describe, Alice's husband liked to be involved with the children, and he also helped out by contributing to the public side of childcare, by offering lifts to other parents with children, rather than do the household chores at home. Alice's husband's was able to 'cherry pick' his contribution to roles at home in accord with Maushart (2003), who makes the point that men helping out with the children draws praise from women but, 'Mum remains the default parent and Dad the back-up' (p. 121), as was the case with Alice, despite her praise of him. Alice's solution (and that of the medics Deidre and Jane in Chapter 6) was to employ paid help, in keeping with the increasing demand since the 1980s for domestic labour, to help families where both partners work (Gregson and Lowe, 1994).

Rosa went a stage further than Angela, Alice and Jane, possibly under-taking a 'gatekeeping' role, as described by Allen and Hawkins (1999), and resisting the input of her husband, though she didn't express it in that way. In this 'gatekeeping' role, Allen and Hawkins (1999) show that 21 per cent of mothers exclude their husbands from undertaking more housework and being more involved in the childcare, as mothers are reluctant to 'relin-quish responsibility for family matters by setting rigid standards, wanting to be ultimately accountable for domestic labour to confirm to others and to herself that she has a valued maternal identity, and expecting that family work is truly a woman's domain' (p. 205).

Allen and Hawkins (1999) also found that women used various reasons for excluding men, including it being easier to do themselves, being con-cerned about what the neighbours think and it being harder for men, but none of these possibilities seemed appropriate for Rosa. Rosa led a complex life 'juggling' her career and her family. A highly motivated healthcare scientist, Rosa found it 'very very difficult and stressful at times' but retained responsibility for the house and for organizing the family because otherwise 'it just wouldn't work'. We continue the section of the dialogue with Rosa from above:

Rosa: Nearly every night of the week one or other of them [children] has a music class or rugby or whatever to which I have to take them or pick them up and I don't think he knows that I do all of that and I don't think he has much concept of what time I actually get home of an evening and it's not very much more before him and then I have to cook the tea and then I start working if I need to do some work . . .
Valerie: Is this something you've brought on yourself do you think?
Rosa: I might have done yes . . .
Valerie: Would you like him to take an equal share?
Rosa: Well I think I'd rather do it my way. I couldn't hand over the responsibil-ity, it just wouldn't work. I would have more stress than I have now. At least now I know it's all well organized and I couldn't let anyone else do it. Maybe we

don't try hard enough when it comes to getting men to do their share but I like
to organize things so I know where I am . . .

As with Ann's statement, 'I like the control' (in Hochschild, 2003, p. 104),
and like Angela and Alice, it is possible that Rosa wanted to control her
home, including both the housework and care of the children, perhaps
because the home was one of the few areas that she was able to be dominant (as suggested by Janeway, 1971) or in a gatekeeping role, as Allen and
Hawkins (1999) suggest. Certainly, Rosa took the responsibility of organizing the house seriously, claiming it was 'well organized' and she 'couldn't
let anyone else do it'. However, an alternative explanation would be that
Rosa anticipated that her husband would organize the house badly (or
not at all) and rather than challenging him over his input into the family
home sought to avoid conflict and stress by doing it herself (see Allen and
Hawkins, 1999; Chandler 1991; Hochschild, 2003; Maushart, 2003; Miller,
1986; Potuchek, 1997).

Unlike Angela and Alice, Rosa had been unable to have a career on
anything like equal terms with her husband. Her role was constructed as
wife and 'm[o]ther' despite striving in her scientific career, which, whilst
highly regarded among her peers, remained in the background at home. It
may be that Rosa's husband delegated the responsibility to her so that she
was working within unwritten guidelines, as Edgell (1980) and Mansfield
and Collard (1988) suggest. In this way, despite struggling to advance her
own career, albeit in secret, Rosa was kept in her metaphorical place away
from the select place that her husband inhabited – the public place of men
(Miller, 1986).

Given Rosa's account of how little her husband contributed to the
household roles, it could be that her claims of wishing to lead the domestic
care agenda were a form of post hoc rationalization in a situation where
Rosa had no alternative other than to take the lead with household chores
and childcare. Her husband appeared uninterested in doing either, and
Rosa had no one who could take over the responsibility. For Rosa, it
seemed that she rationalized her husband's input as a way of protecting
him from the chores so that he could pursue his more important job. To
avoid conflict, to have a clean and tidy home maintained to a high standard, Rosa had no choice but to take on the responsibility herself.

SOME CONCLUSIONS ON SECRET CAREERS

In this chapter, we have made visible (or by 'unmasking', according to
Hacking, 1998, p. 58) how women's role remains that of mother and

homemaker and men's is that of 'breadwinner' (Potuchek, 1997). As well as affecting their lives at home, these constructions influence women's working lives and indicate some of the reasons why women have little control over their careers. In our sample, there was the expectation from most women and men healthcare scientists (with the exception of Alec and Alice) that childcare would be mainly the responsibility of mothers, as Hochschild (2003) and Maushart (2003) show. Most of the interviews provided little indication that men took anything like equal responsibilities with their wives for home and family. Of the women who remained married, many accommodated the requirements of their husbands to avoid conflict and the women supported the men in their careers rather than negotiating on account of their own professional needs.

Furthermore, it seemed that the husbands of these women had little notion about the career aspirations of their wives (all the heterosexual participants in our research who were cohabiting were married), which the wives conducted in secret. It was apparent that the women humoured their husbands rather than negotiating with them about their own careers. The women emphasized the importance of their husbands' careers, rationalizing why their own careers should take second place, quoting their lower earnings, the children's schooling, or putting men's needs above their own. The wives knew their place in the home just as they did at work and in both environments made sure they did not create undue friction. For their husbands, the home was a continuation of their public work environment where women were 'other' or 'm[o]ther', working to keep the men content and avoid conflict. For women too, their home lives were a continuation of their public lives where they knew their place and did not challenge the status quo.

Even where the women considered science critical to their lives, most women accommodated their husbands, and hid their thoughts about their own careers. This could be because the women were uneasy about telling the men of their career ambitions, fearing they were overstepping their 'place', but also this was partly because these men saw their wives as 'other' in career terms, separated by their difference, in a similar vein to the way men in science tend to see women in the science work environment. Although Babcock and Laschever's (2003) research refers to negotiations with potential employers over pay, the women appeared similarly less successful than men in negotiating their positions within marriage. This manifested itself not only with regard to keeping quiet about their careers but also within their homes by the unequal sharing of responsibility for housework and the organization of family commitments. What could be described as 'discrimination' by the husbands was as subtle as the discrimination by male bosses at work and was accommodated by women, often in order to avoid conflict and preserve the relationship.

Married women with or without children who earned more than (or as much as) their partners appeared to hold more bargaining power and had more productive careers than those women with children whose partners earned more. When wives earned less than husbands and there were children, these differential salaries placed a significant restriction on the career development of the women. Despite earning nearly twice the full-time median salary in Great Britain for men and women (according to ONS, 2016), the seven women who earned less than their husbands presented positive views on their husbands' greater earnings, in keeping with the analysis of Potuchek (1997) and Brannen and Moss (1991) who show that wives often downplay their own earnings. Indeed, several women 'talked up' the importance of the career and salary of their husbands (Maushart, 2003, p. 77).

The heterosexual married women healthcare scientists interviewed all took on more responsibility than their husbands for the home, whether for children, husbands or housework. Their husbands did not undertake anything like equal responsibility for, or carry out much actual housework. The women tended to report being pleased with the 'help' gained from husbands (usually for the technological tasks) and accepted men's lesser input as the 'norm'. The reasons that women continue to take on this burden of responsibility could be for a number of reasons. It could be the only place that women could exert power, as Janeway (1971) suggests, or it could be due to the symbolic nature of the responsibility, as Martin (1984) suggests. More likely, we feel, it was because the women had little choice other than to bear the burden of domestic care agendas, seeking to avoid conflict as a way of preserving their marriage rather than (as in Tina's case) risking divorce.

Arguably then, women with or without children accepted responsibility for family matters because they had no alternative option, even though juggling work and home life was manifestly difficult, nor were the husbands offering to take responsibility for the home organization away from their wives. As involved interpreters of the interview texts and from our position within our research, it seemed that there are several other reasons why this might be. First, both women and men expect women to take prime responsibility for the home (albeit possibly delegated). Second, it could be 'lagged adaptation' as Gershuny et al. (1994) describe, where women and men eventually reach a shared understanding of each other's needs, but where husbands get the better deal. Third, as Maushart (2003) describes, men may not carry out the home, house and child work well and women eventually avoid pressurizing their men to improve their houseworking skills and so are left with an uninviting choice between continual conflict or taking on the lion's share themselves.

Most of the women with children in our study demonstrated that they wanted to continue their commitment to science. However, managing a delay in their careers while their children were young without forgoing their careers in the long term was the difficulty. Most of the married women with children took on the brunt of the housework and caring for children, yet kept quiet about the amount of 'second shift' (Hochschild, 2003) they undertook. Those who were able to pay for help in the home seemed, perhaps not surprisingly, to cope better than those who undertook the household and childcare duties themselves. The careers of the women were hindered by unsympathetic husbands or, on at least one occasion, teenage children. As a consequence, women mostly conducted their careers in secret, experiencing private patriarchy at home as well as public patriarchy at work (Walby, 1989).

5. Creative genius in science

Inevitably then, women can only be given minor roles, if any. They need to be kept to well-scripted roles with limited opportunities for improvisation.
(Höpfl and Hornby Atkinson, 2000, p. 139)

THE LACK OF THE FEMALE CREATIVE GENIUS

In this chapter, we note the lack of the female 'creative genius' and explore how and why women are excluded from this accolade to be positioned in the lower echelons of genius and of science. We first review how the label of genius is awarded mainly to men and how creative genius is associated with the masculine form. We then look at how women are usually positioned in support roles doing 'women's work' and subject to the gender schemas described by Valian (2004). We see how those women who develop careers in science have career trajectories that exist within a 'labyrinth' (Eagly and Carli, 2007, p. 6) of barriers that curtail the opportunities that are offered to men rather than women. With reference to the words of our interviewees, we see that even those women who become leading research scientists are not regarded as being brilliant or as creative geniuses by their (usually male) bosses. We also report how the potential of clever young women on the way up is also often ignored.

Within our research sample, we are not attempting to determine whether or not women are or could be geniuses. Our intention is, rather, to articulate how women in science are precluded from accessing opportunities that might enable them to achieve the levels in scientific research required to be regarded and treated as 'creative geniuses'.

While there are limited examples of women gaining roles as research leaders in science, most high-level roles involving notions of being a 'genius' with creative research skills continue to be held by men (Battersby 1989, pp. 2–3). Perhaps the present day lack of women in leading research roles in science is unsurprising giving the enduring positioning of women scientists in supporting roles, women's place in science being, customarily, to facilitate and sustain research undertaken by male scientists.

Genius is closely associated with science and the notion of the creative or scientific genius is a masculine genius who exerts his power and authority over others (Simonton, 1988). The world recognizes scientific genius in a number of ways, the most notable being the award of a Nobel Prize for 'the greatest benefit to mankind' (Nobel Prize, 2017). Of a total of 881 different individuals awarded the Nobel Prize between 1901 and 2016, women have won the prize 49 times. Of these 49, six women were awarded the prize in chemistry or physics compared with 274 men. In 2016 all awards were made to men. Marie Curie won a Nobel Prize twice, being one of four women winners in chemistry and one of two women who won the prize in physics (ibid.).

In the UK, probably the most highly regarded scientific award is that from the Royal Society, where women represent only 9.5 per cent of current Fellows elected (144 of a total of 1512 [Honorary Fellows not Royal Fellows]; Royal Society, 2017). Even in 2017, the percentage of women elected was only a quarter (15 of 61), a smaller percentage than in 2016 (Royal Society, 2017). The first female Fellows were elected in 1945 and Dorothy Hodgkin, who was elected in 1947, is the only woman to have both a Nobel Prize and to be elected as a Fellow to the Royal Society (ibid.).

Men then, rather than women, are awarded respect for brilliance and recognized as the creative genius. Ironically, despite many of the stereotypical characteristics of geniuses being associated with the traditionally female qualities of 'emotion, sensibility, intuition, imagination', women have long been excluded from notions of genius and eminence in science (Battersby, 1989, p. 3). These characteristics of being a genius, imply a 'rhetoric of exclusion [that] praised "feminine" qualities in male creators . . . but claimed females could not – or should not – create' (ibid.).

Battersby (1989) traces the history of genius from Aristotle (who 'had suggested that a woman was a failed biological experiment', p. 119) leading to our modern usage of the term. Of the various 'strands' in the history of genius, the most familiar is the 'potential for eminence' (p. 156). Rousseau dismissed the possibility of women being geniuses in the eighteenth century: 'Women in general possess no . . . genius' (Rousseau, 1758 [2009], translated by Citron, 1986, quoted in Battersby, 1989, p. 36). Around 30 years later, Kant described genius as *'ingenium'*, an innate quality from which most women (but not all) were excluded. Kant considered that some women could become geniuses but they might thereby suffer and become 'ridiculous' or 'loathsome (*ekelhaft*)' (Battersby, 1989, pp. 76–7, quoting Kant, 1790 [1951]). In 1807, in *Letters on the Intellectual and Moral Character of Women*, Duff 'explicitly argued that women can't be great geniuses' (Battersby, 1989, p. 5, citing Duff, 1807). Battersby (1998) argues that social images of genius and masculinity are hard to disentangle because '*Normal* selves are attached to male bodies; *supra-normal* selves are also attached to male bodies; women, by contrast are credited with *abnormal* selves' (p. 21; emphasis in original).

Battersby's arguments are aligned with observations made by Annandale and Clark (1996) who put forward the view that male bodies have, within medical sciences, been idealized, associated with normally good health and unconstrained intellectual abilities. By comparison, despite lack of evidence for such claims, women's bodies are treated as 'other', frail and unreliable (especially given their potential for maternity, as discussed in Chapter 6). Despite a lack of evidence demonstrating women's physical fragility, assumptions about compromised female health, combined with women's apparent capacity for reproduction, extend into organizational contexts and are constructed as precluding women from being identified as ideal worker material (Witz, 2000). Problematically, organizational views about women's fragile health affect judgements about their intellectual capabilities. Women's supposedly poor health is often presumed to compromise female intellectual capacity to think and act rationally, creatively yet incisively (Gatrell, 2011a; Witz, 2000). One consequence of these negative views regarding female talent is the exclusion of women from attaining the status of genius.

Genius, Battersby (1989) points out, is not a factual but a 'descriptive' or 'complex value-judgement', which, like all evaluative judgements, is 'normative' [by] 'telling others (and oneself) what the response *ought* to be' (p. 124; emphasis in original). Genius is reserved for the public place of men. However, not everyone sees genius as 'descriptive'. Influential attempts at empirical determinations of the potential for genius originated in the nineteenth century by the work of Francis Galton, as Battersby (1989) describes: 'In general, scientists (including social scientists) and psychologists who have gone for an "empirical" study of genius, originality and creativity have tried to play down the evaluative elements in such judgements' (p. 125).

In 1869, Galton (published *Hereditary Genius*, a book that influenced eugenics and many subsequent tests for measuring the 'intelligence quotient' (IQ) including the UK 11-plus examination (Galton, 1869 [1972], cited in Battersby, 1989). Galton devised ways of 'measuring' genius to demonstrate its relationship to heredity, which Battersby (1989) suggests Galton was keen to emphasize, being a younger cousin of Darwin.

The reputation of IQ tests been brought into disrepute as they were based on 'evidence of systematic fraud' in the research undertaken by Cyril Burt who promulgated their introduction in the first half of the twentieth century (Parrington, 1996, p. 3). They have also been shown to be biased but versions of the IQ test continue to be used in the UK in the 11-plus examination and are subject to the same criticisms (Parrington, 1996).

Historically, quotas for 11-plus pass rates limited the pass rates for girls but was 'concealed from public debate' (Pennell and West, 2003, p. 53, citing Gallagher, 1997). Gallagher's review reports on the 'capping' of places for girls who were deemed to have qualified for places in grammar schools as a result of their performance in the 11-plus test' and 'notes the main justification for this policy was that boys performed less well than girls at 11 years of age because they matured at a slower rate than girls' (ibid.). Middle class bias, month of birth have also been shown to affect 11-plus outcomes (Hart et al., 2012), favouring those who understand the cultural context of the tests (Parrington, 1996). The Newsom Report in 1963 (quoted in Pennell and West, 2003, p. 51) notes: 'The girl may come to the science lesson with a less eager curiosity than the boy'. These biases in the education system are likely to have affected the women in our study who were at school prior to 1980.

More recently, according to the Lynn and Kanazawa (2011), there are negligible gender differences in the performance of pupils in IQ tests so girls should be just as likely as boys to be 'gifted' or to become a 'creative genius'. Reis (2003) similarly describes the 'male conception of gifted-ness. If girls were identified, they were thought to have the mind of a boy as they could not be both gifted and feminine'. Reis (2003) records how Annemarie Roeper's article 'The young gifted girl' in the first issue of the *Roeper Review* in 1978, was one of the first research projects to focus attention on talented young women (as opposed to young men). Since this first edition, the *Roeper Review* (which includes all gifted children, not just girls) has addressed the development of young women to recognize their gifts and to emphasize their special talents.

Arguably, 'creative genius' isn't a status aspired to by an individual, but a position controlled by others, usually men, who identify men as having that potential position. Women tend to remain invisible in the science world, where the example of the expert is male. Historically, it has been seen as inappropriate for women to be experts, particularly if it meant women could influence men (Rossiter, 1982) and women were excluded from authoritative positions in science 'on grounds of propriety' (Code, 1991, p. 226). Smith (1974) also argues that because of the inequality of knowledge and experience, men impose their thoughts and concepts upon the world and upon women: '[T]he two worlds and the two bases of knowl-edge and experience don't stand in an equal relation. The world as it is constituted by men stands in authority over that of women' (p. 7).

Historically, women's knowledge was seen as 'naturally subjective' and using this reasoning women are precluded from the state of being knowledgeable (Code, 1991, p. 10). Code (1991) explores how knowledge is related to gender and sex and how women are excluded from having knowledge. Some feminists, including Code (1991) and Harding (1991) consider that there are ways of knowing that are distinctly related to being female and dependent on women's different experiences, but Code warns that women may concentrate too much on experience to the detriment of discovering knowledge. She also argues that the qualities associated with women are not valued as knowledge by men mainly because women's knowledge is experiential:

> Some feminist theorists have maintained that there are distinctively female – or feminine – ways of knowing; neglected ways, from which the label 'knowledge', traditionally is withheld. Many claim that a recognition of these 'ways of knowing' should prompt the development of new, rival or even separate episte-mologies. Others have adopted Mary O'Brien's (1981) brilliant characterization of mainstream epistemology as 'malestream', claiming that one of the principal manifestations of its hegemony is its suppression of female – or 'feminine' knowledge. (Code, 1991, pp. 12–13)

Code (1991) suggests that although knowledge is usually considered to be objective, requiring theoretical analysis, knowledge is also subjective because it is produced 'by the processes of its construction by specifically located subjects' (p. 255). She highlights that the continued use of stereo-types demeans women's abilities to think, to be knowledgeable and to have authority and they are therefore afforded little credibility (p. 223):

> [C]omplex structural patterns converge to contain women within undervalued cognitive domains and to thwart their efforts to gain recognition as fully author-itative members of epistemic communities [including] the tenacious cluster of stereotypes that underpin and reinforce sociocultural representations of women as scatterbrained, illogical, highly emotional creatures, incapable of abstract intellectual thought.

Observing how genius and expertise are associated with knowledge and masculinity, Code (ibid., citing Baumgart, 1985) notes 'the curious distinction between knowledge and experience', where '[a]ccording to the stereotypes, women have access *only* to experience, hence not to the stuff of which knowledge is made' (emphasis in original). She highlights the double standard of expertise and the bias involved between men's pro-fessional knowledge as doctors and women's experiential knowledge as nurses, commenting on the legal case reviewing infant deaths in Toronto's Hospital for sick children:

> Alice Baumgart observes, 'When lawyers, who were mostly men, questioned doctors, the questions were phrased in terms of what they knew. When nurses were on the stand, the question was, "based on your *experience*. . .". Experience

in our society is considered second class compared to knowledge. Nurses should not know'. (Code, 1991, p. 222, quoting Baumgart, 1985; emphasis in original)

An early (and continuing) example of how women's knowledge and experience was constructed to the detriment of women is in midwifery, which is traditionally regarded as the domain of women but became medicalized by men who introduced technical expertise with the introduction of forceps (Fara, 2004). Women who were midwives did not use technology, in contrast to the man-midwives who brought instruments with their new trade, which they used instead of hands to deliver babies. The difference between the male and female role in midwifery is illustrated in Fara's book *Pandora's Breeches* (2004) where the 'man-mid-wife' stands in a laboratory with instruments and a pestle and mortar, whereas the female midwife stands by an open range in a room, probably a kitchen, with patterned floor covering (Fara, 2004, p. 28, Figure 6), emphasizing the notion of women's place in the home. 'Man-midwives' often charged for their services and had an advantage in being able to establish medical colleges (from which women midwives were excluded). Male midwives later evolved into obstetricians (Schiebinger, 1989), positioning themselves in the place of power and 'action' (Miller, 1986, p. 75).

Today, barriers remain between midwives (usually women) and medically qualified obstetricians (usually men), the only professional group allowed to use instruments to deliver babies. When births are not going easily, the (often male) obstetrician is called upon to be the 'expert' and to intervene with technology to deliver the baby safely. Women remain in their place in the less expert role – their knowledge is denied. Midwives are refused access to the professional masculine group that is assigned by society as having the superior knowledge and expertise, able to be called on in an emergency, as symbolic as a knight on a white horse.

Haste (1993), in a study of undergraduates and school students, found a 'clear relationship between subjects perceived as scientific and subjects perceived as masculine' (p. 77) and where science is seen as 'complex or difficult (rather than simple)' (p. 78). Haste (1993) suggests that '[t]he relationship is between science as an activity and a form of knowledge, and culturally defined masculinity', concluding that 'there is no such thing as "female knowledge"' (ibid.). We see in our study that women are disadvantaged in the science work 'place' because they are associated with experience, and not knowledge or expertise.

Such associations between masculinity and expertise affected the women scientists in our study in their relationships with male colleagues and bosses. Below, we look at women's work and careers and explore how women's potential as a creative genius or brilliant scientist is disregarded throughout their working lives.

WOMEN'S WORK

We now consider in relation to science more broadly, how the work opportunities open to women result in their being diverted from the paths most
likely to lead to a distinguished career in scientific research. It continues
to be evidenced how women face and overcome more hurdles than men
to make progress in their careers to overcome the disadvantage of their
gender (Davidson and Cooper, 1992; Edwards and Wajcman, 2005; Höpfl
and Hornby Atkinson, 2000): the gendered division of labour continues to
be central to the way both women and men behave and the roles they play
(Acker, 1990, 1992, 1998; Hearn, 1987; Sheppard, 1989).

Employment of women is highly concentrated by occupation: nearly
two-thirds of women work in 12 occupations (out of 77 recognized occupations) and these tend to be in the five Cs – catering, cleaning, caring,
clerical and cashiering (Equality and Human Rights Commission [EHRC],
2009, p. 3), which are regarded as 'women's work' and where the pay is low,
as Office for National Statistics (2013) data show: 'Men tended to work in
professional occupations associated with higher levels of pay. For example
programmers and software development professionals earned £20.02 per
hour (excluding overtime) while nurses earned on average £16.61 according to the 2012 Annual Survey of Hours and Earnings' (ONS, 2013, p. 11).

It could be argued that women's work may be defined now in the same
manner as in 1893 when Gissing wrote: 'A womanly occupation means,
practically, an occupation that a man disdains' (Gissing, 1893 [1980],
p. 135, quoted in Delamont, 2001a, p. 84). Because less value is placed on
women's work, a gender divide has arisen based on the types of jobs that
women and men do (Cross and Bagilhole, 2002; Lupton, 2000; Women
and Work Commission [WWC], 2006a). Often gender inequalities appear
to be ignored. Nutley et al. (2002) suggest that both women and men
have an 'aversion to recognizing difference and inequality' and 'appear
to be comfortable in turning a "blind eye" to gender issues' (p. 4). Valian
(2004) terms such working environments as 'gender schemas' where both
male and female workers overrate the input of men and underrate that of
women (p. 208).

One example of a woman subject to the gender schemas of her organization and whose potential was not put to the test was found in the interview
text from Zoe, who was in her early twenties and had worked in her public
sector institute for just over three years:

> I pretty much knew within a year that I didn't want to stay here because I didn't
> think that there was any way of me being able to move up into what I wanted
> to do. I thought that I could get experience of lab work and sort of start at

the bottom and work my way up but I'm not sure that you can really do that anymore. (Zoe)

Zoe's potential was not encouraged and a year later she moved to another institute where she hoped there might be more opportunities.

Although massive changes have taken place during the second half of the twentieth century and into the twenty-first, examples of the exclusion of women in areas of paid work are still easy to find. Women are in the minority at the top of nearly all professions: the judiciary, government, medicine, science, management, education and more (Greenfield, 2002a). The Church of England has made a modicum of progress, with the Synod voting in November 1992 to allow women priests to be ordained (Church of England, 2016) but were prohibited from becoming bishops more than 20 years later in November 2014, with the first woman bishop being appointed in January 2015 (ibid.).

According to ONS (2013), about 90 per cent of men of working age with children in the UK are in employment compared with around 73 per cent of working age women. The *Full Report – Women in the Labour Market* (ONS, 2013) notes that the increase in women in the labour force is due to an increase in the numbers of working mothers. Thirty-five per cent of managers, directors and senior officials in the UK are women, which is one of the highest rates in the European Union (EU) (ONS, 2013, p. 12, figures for April–June 2013). The proportion of women in non-director posts on the FTSE 100 boards reached 23.5 per cent in 2015 although the proportion of female directorships is much lower at 8.6 per cent; the FTSE 250 board figures for women are lower at 18 per cent and 4.6 per cent respectively (Vinnicombe et al., 2015).

So what does this mean for women who become qualified in science, and how are their chances of becoming scientific leaders affected? The minutes from the Meeting of the OECD Council at Ministerial Level which took place in Paris in May 2012, notes that:

> [. . .] girls are still less likely to choose scientific and technological fields of study, and even when they do, they are less likely to take up a career in these fields – a concern given skills shortages in the workplace, the generally more promising career and earnings prospects in these fields, and the likelihood of positive spillovers from more skilled workers in these fields to innovation and growth. (OECD, 2012, p. 4)

This report recognizes that women are disadvantaged as they climb the career ladder:

> But irrespective of family commitments many female professionals find it difficult to climb the career ladder. In fact, inequalities increase the higher up the

pay scale you go, so that while on average in OECD countries women earn 16 per cent less than men, female top-earners are paid on average 21 per cent less than their male counterparts. This suggests the presence of a so-called 'glass ceiling'. Women are also disadvantaged when it comes to decision-making responsibilities and senior management positions; by the time you get to the boardroom, there are only 10 women for every 100 men. (OECD, 2012, p. 5)

As Wajcman (1991, p. 5) argues, the historically embedded dominance of men within science was justified (and subsequently reinforced) by masculine discourses, in that science:

> [. . .]was fundamentally based on the masculine projects of reason and objectivity. They characterized the conceptual dichotomizing central to scientific thought and to Western philosophy in general, as distinctly masculine. Culture vs. subjectivity, the public realm vs. the private realm – in each dichotomy the former must dominate the latter and the latter in each case seems to be systematically associated with the feminine.

Women are 'pushed' according to Bagilhole et al. (2008, p. 38) 'into "softer" areas within SET (science, engineering and technology) occupations that are deemed more suitable for women, but which often afford lesser opportunities for career advancement'. Thus, many scientists with PhDs need to find employment in fields other than those for which their training prepared them (Elves and Gibson, 2013) and when women do stay in science they earn less than their male counterparts (Palmer and Yandell, 2013). Thus, women are disadvantaged from the beginning of their working lives, being positioned in lower-status jobs.

WOMEN'S CAREERS

We have already offered explanations for why women may face barriers to being recognized as creative research scientists (if it occurs at all that such prestigious roles might be open to them). We now explore women's career trajectories, suggesting that these are less straightforward than men's, going on to describe how the women in our research developed (or were stalled) in their advancement as scientists.

Opinions differ on why advancement of women's careers is constrained compared with that of men. Social economist Catherine Hakim claims that women (especially mothers) are less career oriented than men (Hakim, 2011). Hakim argues that women in heterosexual relationships are more likely to settle for 'jobs' that can accommodate family needs rather than pursuing a 'career'. Hakim presents this as a proactive and unimpeded choice among women who supposedly lack commitment to paid work. She takes little account of the cultural and gendered organizational and social contexts and practices that, we argue, constrain women's choices. Hakim's view is also in contrast to the research by Blair-Loy (2003) and Wajcman (1998) who show that women can be highly committed to their paid work even when they work part-time and is also challenged by Gatrell (2005) and others.

Despite an increase in the number of women in professional and managerial careers, the careers of women are regarded as less well structured and less planned than men's (Burke, 2002; Dyhouse, 2006; Höpfl and Hornby Atkinson, 2000). There was relatively little unemployment in the 1960s and young women (and men) tended to leave school to enter employment rather than go to university. Women tended to enter occupations that might develop into careers, as nurses, secretaries, to work in banks or to take on other vocational occupations (Dyhouse, 2006), including working in healthcare science laboratories, as Valerie did. Dyhouse (2006) reports that other changes took place from 1970 onwards that influenced women's career aspirations, including more sexual freedom due to the 'pill' and legalization of abortion, the prominence of feminism, and a decline in the recruitment of teachers, which encouraged women to move into other areas of work.

Alongside the discourse of the masculinity of science is the discourse of the masculine career, where 'directed rationality ... drives the pursuit of career objectives' is the accepted route (Höpfl and Hornby Atkinson, 2000) and where careerism is one of five masculine identities[1] – a particular form of competitiveness associated with 'hierarchical advance' embedded into organizational practices, according to Collinson and Hearn (1994, pp. 13–15). In addition, the 'notion of career "success" is traditionally defined in what are male terms', that is, salary, power, status' (Höpfl and Hornby Atkinson, 2000, p. 130). Early career models were based on men's pursuit of objectives and limited thought was given to how women experienced careers.

The manner in which career advancement is achieved is changing and evolving, being influenced by psychology, sociology, education and management (Peiperl and Arthur, 2000) but is always rooted in the idea of choice, where 'the commonsense view of career [is] as trajectory, as a strategic life plan' (Höpfl and Hornby Atkinson, 2000, p. 134). Höpfl and Hornby Atkinson (2000) argue that women may reject traditional career trajectories because of 'competing demands, and the character and value of rewards' (ibid.). They also suggest that 'women have long been aware of the problems associated with the unitary notion of organizational commitment' and that women's 'duality of commitment to home and work' leads to 'ambivalence that is at the root of women's engagement with work' (Höpfl and Hornby Atkinson, 2000, p. 140).

Although it has been argued that career hierarchies may be disappearing in the corporate world (Littleton et al., 2000), with portfolio careers becoming increasingly the model, the traditional career hierarchy is still very much in evidence within scientific organizations in both public and private sectors. Scientists work in a 'realist' world, where notions of scientific 'objectivity' and 'rationality' prevail and our research shows how traditional career paths remain the norm for men. Kirkup et al. (2010, p. 8) point out that women's chances of advancing in a research career in science are less than for men:

> Men are more likely than women to take up SET management positions (37.7 per cent of all male SET professionals/associate professionals compared to 28.6 per cent of women in the same occupational group). But a larger proportion of women work in (lower level) SET associate professions: as technicians, draughtspersons and inspectors . . . (26.5 per cent among women compared to 16.5 per cent among men).

Science is driven by the need to employ highly qualified scientists, some of whom will progress to direct their own work programmes, with a few becoming scientific research leaders, perhaps accomplishing board-level status. Fewer women than men, however, will progress to this high level and few, if any, are identified as creative geniuses.

One possible explanation for the lack of promotion among women scientists lies in the propensity for men to identify other men, rather than women, as potential or actual creative geniuses (see, for example, our earlier note on the lack of women Fellows of the Royal Society). This disadvantages women because, as Babcock and Laschever (2003) describe, in a study of school children, those classified as having potential may be singled out by leaders who then invest in their development. Because women are identified less often than men as having potential, they do not have the chance to benefit from the support necessary to advance their careers: 'Inevitably then, women can only be given minor roles, if any. They need to be kept to well-scripted roles with limited opportunities for improvisation' (Höpfl and Hornby Atkinson 2000, p. 139).

Women are thus restrained by structural barriers and the expectations by male managers of how they view women's potential: they are not offered the same opportunities as men and are unable to make choices because they do not know what is possible. Women may also be seen as a threat because they do not fit the norms of the masculine science world: they challenge the 'corporate and male definition of the situation [where] . . . issues of personal responsibilities, and personal meaning are thrown into focus' (ibid.).

The route through a career and what is regarded as a successful career is therefore likely to be different for women and men (Höpfl and Hornby Atkinson, 2000; White et al., 1992). Höpfl and Hornby Atkinson (2000) and Haste (1993) suggest that women's careers are based on experiences rather than being rationally directed. Whilst experience may be an important factor for women overall, scientific careers are rooted in vocational choice for both women and men, as was the case for the women interviewees both from the public and private sectors, many of whom were passionate about science and their individual subject (Mallon et al., 2005).

Nutley et al. (2002) indicate that the continuing lack of women in senior management positions is a consequence of women having children: the careers of women are affected more than those of men as the man's career usually takes precedence in a relationship. They comment that women may be more willing to take the lesser role because women's careers are less important to women than men's careers are to men and women do not 'define' themselves by their careers (p. 4). However, as we show in the next chapter on motherhood, women's social identity may be more closely related to their occupation/career than to being a mother, and motherhood may not fulfil all that women want in terms of a challenge (Brannen and Moss, 1991; see also Blair-Loy, 2003; Gatrell, 2005; Potuchek, 1997). Furthermore, not all women scientists have children, yet the opportunities for all women scientists to be recognized as 'creative' research geniuses remain limited.

Eagly and Carli (2007) describe a 'labyrinth [which] contains numerous barriers, some subtle and others quite obvious, such as the expectation that mothers will provide the lion's share of childcare' (p. 6). The labyrinth to be navigated hinders the progress of women through career structures: where advancement of men is straightforward, women are obliged to circumnavigate twists and turns. Even in professions such as nursing where there are many more women than men, male nurses progress faster despite having less appropriate qualifications than women (Davies and Rosser, 1986). Hearn (1987, p. 128) describes the 'patriarchal feminine' role of the paramedical semi-professions that complements the 'masculine' doctor, and the example of such control by the profession over nursing and radiography is reviewed by several authors (Davies, 1995; Davies and Rosser 1986; Witz, 1992) who research the low status of the nursing and radiography professions. A similar relationship is manifested in clinical diagnostic and other healthcare science laboratories where biomedical scientists work within a hierarchy where the powerful medical profession manifests its professional expertise and exerts controls over the less powerful biomedical science, which could be considered a 'semi-profession'.

WOMEN IN HEALTHCARE SCIENCE

Kerckhoff (1995) has related the importance of 'linkages' in relation to careers, between family and school, between school and work and within institutions and contends that although 'individual differences are bound to alter the odds at the individual level', the most significant linkages are to be found at institutional level (p. 342). Kerckhoff (1995, p. 341) also illustrates how different kinds of schools and institutions control mobility, so that some people make progress by accessing opportunities at work and others are held back:

> [The] literature suggests that there are two general types of societal patterns of linkage between labour force careers and the form of the education system. In the first type, the education system awards highly specific, occupationally relevant credentials and encourages specialization beginning at relatively early ages, well before labour force entry ... and they provide a basis for orderly career patterns not wholly dependent on the opportunity structure in particular firms ... In contrast, other educational systems award very general, nonspecific credentials that provide their students with little direction as they enter the labour force ... Early work-career patterns are less orderly, and they are more dependent on the opportunities provided by workers' current employers.

These statements are highly relevant to careers in healthcare science where both routes are used to access careers. The first route is highly relevant in a healthcare science career in biomedical science where a specific education system has been developed and controlled by the professional institute, the Institute of Biomedical Science (IBMS), although the Honours degree qualifications are undertaken within universities. Within the IBMS career structure, biomedical scientists undertake statutory registration with the Health and Care Professions Council (HCPC) and take further professional exams as they develop their careers in clinical diagnostic pathology laboratories (rather than in research laboratories). Clinical scientists usually start work with a PhD and take a different career route, usually in research or clinical diagnostic pathology laboratories (see note at end of Chapter 1).

Within the first route defined by Kerckhoff (1995), it is difficult to move from one career path to another without starting on a different career path from scratch. For biomedical scientists, the specificity of the biomedical science career structure and the professional barriers in place make it particularly difficult for those biomedical scientists who wish to move to a designated research post or to a more clinical role as a clinical scientist. Despite there being considerable overlap in knowledge, experience and responsibilities expected of the two professional groups (biomedical and clinical scientists), moving from one career path as a biomedical scientist to that of a clinical scientist presented significant problems for the women in our study.

The second route highlighted by Kerckhoff (1995) is found in public sector organizations such as the National Health Service (NHS) and Health Protection Agency (now Public Health England), where post-degree qualifications including PhDs and MScs may be undertaken at the place of work through part-time study. It is also the path used by biomedical scientists who wish to obtain more widely recognized postgraduate degrees than the qualifications attained through their professional institute. Although research scientists may start off as biomedical scientists, the most straightforward route to becoming a research healthcare scientist is to have a specialist science education, including a good science first degree followed by postgraduate qualifications, usually a Masters degree followed by a PhD. Other routes are possible for those who start work without a first degree. However, study for academic science qualifications whilst in paid employment is a route that tends to take longer and be less straight-forward than for those who obtain their postgraduate degrees before they start work in paid employment.

In our research, both routes were used by our participants covering a wide range of academic achievements gained over various periods of time (see Appendix for more details of participants). In our study, about a third of participants did not begin their careers with a university education but obtained professional exams or degrees by part-time courses where study was outside their normal working hours, usually for IBMS qualifications or Masters degrees. The women who studied part-time to take profes-sional, graduate and postgraduate qualifications show how difficult it can be for women to move from a practitioner laboratory role to a more scien-tific role and career.

Most of the women who left school at 16 or 18 initially progressed through the grades of biomedical science and some later aspired to be research or clinical scientists. Some moved into laboratory management posts (the usual career progression route for biomedical scientists) before realigning themselves to make research their highest priority as Ella and Paula did, described elsewhere in this book. Several women found their progress as research scientists impossible and left science altogether.

Below, we show how women in our sample, even those women who became very senior scientists recognized by receiving chairs at prestigious universities, were apparently never considered as being creative geniuses. Similarly, the potential brilliance of some of the younger women was not noticed or was ignored and they were kept in their place in the lower ranks of science because they didn't fit the expectations ascribed to the high-flying masculine scientist – they were 'other' and invisible. As noted earlier, we do not claim that all our women research participants would necessarily have attained the status of creative genius, but we do suggest that they were often denied the opportunities that might have facilitated this.

In particular, we show how women's careers in science may be dominated by the presence (or otherwise) of a senior male sponsor who is usually their boss. Without such sponsorship, even if this is lukewarm, the potential for women's career advancement is limited. We show how even where potential for advancement was spotted early on in women's careers by male bosses who, by their sponsorship, helped the women achieve leading roles in science against the odds, they were often unable to sustain such advancement. Lack of continuing sponsorship seemed to take women away from research science as senior women scientists became thwarted again later in their careers by repeated professional barriers and structures. Even though Carolyn, for instance, reached a pinnacle where she was able to influence at board level, to achieve this advancement she moved away from science into high-level management and left her scientific interests behind.

COULD SHE BE A FEMALE GENIUS?

We look first at two women (Ella and Diana) who reached the pinnacle of their careers by being awarded academic chairs in major UK universities despite leaving school at the age of 16. Neither woman came from an academic family but each had a fascination for her subject. Ella left school in the 1970s with O Levels (General Certificate of Education Ordinary Level) and became a registered biomedical scientist in a clinical diagnostic laboratory, gaining professional examinations, working her way up through laboratory science into laboratory management, and working some of the time in research posts. She obtained a Masters degree by part-time evening study. In this laboratory, her aptitude for research was noticed by her medical boss, 'Philip', probably not seeing her as a creative genius, but seeing a potential scientist. He spotted potential in her that she thought included 'always questioning', showing initiative, picking up possible projects, and being prepared to work late on these ideas that were in addition to the usual day's activities. Diana also obtained professional qualifications with the IBMS but had always worked in R&D and had never been a laboratory manager.

Ella was married and Diana was single and neither had children. We now look in more detail at the careers of these two high-achieving women and the barriers they met and overcame; we then consider the barriers they faced in their later careers, which they were not able to overcome, despite having prestigious roles in research and having international reputations. We do not claim (as indicated earlier) to judge whether or not these women scientists had the capacity to be creative geniuses, rather we make the point that their opportunities to reach the top echelons of scientific research were blocked on several occasions from an early age.

Neither woman was seen as a creative genius by family, school or bosses although each had a male boss who spotted potential in them and encouraged them, at least for a time. Each had to overcome many barriers to reach their positions of distinction. Both women built research teams and held academic appointments whilst working in the public sector, though they were not, nor did not they see themselves, as typical academics and were focused on the clinical application of their research. Over time they developed international reputations and became internationally renowned research healthcare scientists, each specializing in different areas of research.

As a senior practitioner in her healthcare science diagnostic pathology laboratory (where clinical specimens are processed) rather than on a scientific scale, Ella undertook many diverse work responsibilities. The more involved she became in research, the more interested she became and research became the focus of her work. Although Ella was supervising the research of others, she did not initially understand that to be recognized as an authentic researcher she needed the additional qualification of a PhD:

> I didn't see why I couldn't do research without doing a PhD and 'Philip' said that if I didn't do a PhD I'd never do unsupervised research or write my own grants. Of course he was right but we never really discussed much about my career development and I felt I was being pushed into it but I didn't like to show my ignorance by asking. (Ella)

Ella was ambivalent about working for a PhD but eventually understood that Philip was encouraging her to develop a research career by studying for a PhD, which would benefit her, though he never discussed this with her as part of her career development. Philip also expected Ella to write his grant applications but he didn't describe this as a means of developing her skills and instead she interpreted his input as a way of him coercing her to write them, a necessary and important task but one that she considered he didn't want to do himself.

Ella was also expected by Philip to continue managing the diagnostic laboratory with some 50 staff in it as well as the research laboratory. So although Philip encouraged Ella to study for a PhD, problematically he did not offer her relief from workload or flexibility at work:

> So I ran the research department which had quite a few people in it and I was also the lab manager in the diagnostic lab so I basically looked after everything in the department and nobody bothered about the fact that I was trying to develop a research career and run the laboratory at the same time. (Ella)

So Philip identified potential in Ella but offered limited support and only on his terms. Being knowledgeable and confident in himself as a medically qualified and respected researcher, Philip seemed unaware of the extra support Ella needed to progress, especially as she came from a home where academic qualifications were virtually unknown.

Some years later in the same post, Ella achieved her PhD by part-time study. The backing of Ella's boss, without which she would not have been able to pursue a PhD, was shown again to be circumscribed. He set the terms of his support and defined the parameters of his help, which did not extend to helping her receive the recognition of a full academic post and she was placed on an 'academically related scale' with consequent lower pay:

> So it was decided that I should have an academic post on the academically related scale, not a full academic post. I was not happy about this but I really wasn't given much choice. (Ella)

There would have been significant financial and status advantages for Ella to have been given a full academic post, but this was denied her at this stage. She was not able to negotiate or challenge it effectively with Philip, as Bowles et al. (2005) and Babcock and Laschever (2003) describe that women not only negotiate less well than men, but they seem to be offered fewer opportunities than men to negotiate. Among women and men managers, some regard it as being somehow unseemly for women to negotiate for higher status or higher pay. Consequently, those who do attempt such negotiations status may experience conflict and unfavourable outcomes.

In successful negotiations, women and men move towards shared understanding of language and 'shared assumptions and beliefs' (Haste, 1993, p. 25). Yet persuading others requires negotiation skills, which is gendered in favour of men (Babcock and Laschever, 2003; Bowles et al., 2005). Women are constrained from negotiating – if women ask for more, they are condemned for behaviour unbefitting a woman. Babcock and Laschever (2003) describe this phenomenon as 'Nice girls don't ask' (p. 62). Women in science and other disciplines avoid negotiations around pay and prospects because they have been 'socialized from an early age not to promote their own interests and instead to focus on the needs of others' (Babcock et al., 2003, p. 14). Typically, Ella did not tackle her boss but continued in her position.

Thus, Ella was not given a choice and she felt thrusted into becoming an academic without being enabled in this course of action by her medical boss. On the one hand he encouraged her to do a PhD but on the other he did not facilitate a reduction in her other work commitments. Ella was expected to retain her management responsibilities without the recognition of becoming the laboratory manager and she was expected to pursue a research career but without the recognition of being a research scientist. She was seen as the caring and compliant woman, not the high-flying researcher with innovative ideas of her own with the possible potential of being a creative genius.

When women realize they have an academic interest in science and want to study further to progress, they need the support of a medical or sometimes senior scientist boss to sponsor them to help them progress. This support includes encouragement to find a suitable PhD project and supervisor, and if a more clinical role is sought, in accessing the training and examinations of the Royal College of Pathologists (RC Path). However, supporting women biomedical scientists in their careers is not part of the normal responsibilities for medical staff and may be resisted. This is where the role of the sponsor is crucial. In a lecture in 2014, the Chief Medical Officer, Dame Professor Sally Davies, said it was important for a woman scientist to identify a sponsor to be an advocate for her. So it is apparent that even in the twenty-first century, many women need to find a sponsor or advocate to be part of the equivalent of an old boy network in order to succeed in a career in science. Similarly, Hollenshead (2003) recommends '[a]dvocacy and coalition building' as well as '[l]eadership and support from the top', as two (of six) 'components of success' (p. 223). Women scientists seem unable to advance alone on their own merits. Mentoring and advocacy schemes were not widely available in the organizations in this study and it seemed that women seeking to pursue a research career had to rely on being 'spotted' – for the unassuming woman this proved difficult.

To advance from a practitioner role in science to leading a research team, to progress to a decision-making role in a high-level post means overcoming massive barriers that women rarely overcome. It also means identifying and accessing opportunities that may not be obvious and the sponsor or advocate is needed to encourage and open doors to help a young scientist to progress. But access to opportunities may also be subtly and not so subtly discouraged by hegemonic bosses and initial encouragement and support may be curbed and short-lived.

Ella subsequently gained a full-time academic post as a senior lecturer directing her own work programmes and overcame even more barriers, not least at her interview to become a professor in a leading university:

> I wasn't expected to get the chair because I didn't have the right background for
> a start. I didn't have a major programme grant and all the work I've done has
> been clinically related, which has been difficult to fund. I've also always done
> a lot of teaching, which isn't necessarily an advantage for how an academic is
> regarded. It wasn't thought that I would get it and also because I'm a woman
> and it was J University. (Ella)

Ella continued:

> It was the most frightening interview I've ever been to. There were about eight
> of them, most of them non-scientists but somehow or other I did get it. They
> asked me about the way I'd come up and you know what research I'd been doing
> and what teaching I'd done and things like that, but it was really the manner in
> which it was conducted and it was very confrontational. They were actually very
> aggressive, verbally quite aggressive. (Ella)

Later still, Ella moved from her full-time academic post and gained
a senior management post in a public sector institute. But even here her
position was not secure. She climbed the slippery career ladder in research
but in the later stages of her career her fortunes took a plunge, which she
was powerless to prevent. Changes in the management structure in her
organization made her vulnerable and she lost her position as a high-flying
scientist with clinical responsibilities and was effectively demoted by being
slotted into a hierarchy with a new medical boss. Ella felt demotivated:

> It is very difficult here now. In fact sometimes I hate my job here and if it wasn't
> for my supportive staff who keep me going, I think I'd go. Apart from them I
> get a lot of support from international colleagues. Our work together is very
> satisfying. (Ella)

Following up Ella's career three years later, Ella was indeed preparing for
early retirement. She had given up the struggle.

Was Ella able to make choices? Ella's career path was not linear: it was
more of a labyrinth than an upward trajectory (Eagly and Carli, 2007,
p. 6). Like many women she spent her working life 'careering' in a reactive
way rather than following a planned career path as her work choices were
limited and a long-term career strategy had never been a possibility. She
reflected on the barriers she had met:

> I think it's actually quite difficult to know whether the barriers were because I
> was a woman, or because I came up through the biomedical science route but I
> think it is both. (Ella)

Arguably, Ella did exercise choice but endured a science world of masculinist, structural and professional demarcation barriers. She also had the

disadvantage of leaving school early, as Archer (2000) describes, and came from a family with no academic background. The equivocality of choice is explored by Archer (2000, p. 69), who points out that we are bound by our circumstances that limit us when we try to make choices:

> Necessarily this means that agents must know their preferences for experiences which they have not yet tried and which might alter them considerably if they did so, for relationships which they have not yet undergone, but which, like motherhood, there is no going back upon, and for ways of living which they have not sampled, but which, like early school leaving, will alter their life-chances if they do.

Over time, Ella seemed to change from being passive to becoming a much more active chooser (Archer, 2000) and she took positive actions to progress her research and her career. She applied for senior academic posts and overcame considerable gender, professional and structural barriers to reach the academic heights of a highly regarded research scientist and professor in a high-ranking university. She commented, however: 'I never meant to be here in this position': even when she was in a position of scientific influence, she lacked self-assurance. So what chance would Ella have of becoming recognized as a creative genius? Following 'Philip's' support in the early stages of her career, she never again had anyone who recognized and encouraged her potential; she had a degree of success in masculinist terms and was respected in her field but, unlike Philip, had little confidence, and in the later stages of her career became invisible, recognized as 'other' – an outsider.

For Ella then, although the support from Philip helped and protected her in the early stages of her career, it also curtailed her progress when he placed her on an 'academically related scale' and did not recommend her for a full academic post, which meant a lower salary than she was entitled to. It seemed he was prepared to encourage her but only to a limited extent. As soon as money became an issue, when he was required to challenge the system that was outside his area of direct control, he didn't fight on her behalf. Whilst Ella's boss did not withdraw his encouragement, the support was on his terms.

Ella was thus embodied by her difference as the invisible person who cared, served and organized not only her boss but the laboratory as well, even giving up her own time to ensure the laboratory ran smoothly. Ella knew her place and did not attempt to negotiate for a change in pay scales. She expected her boss to notice the additional hard work she was undertaking on his behalf (much of it in her own time) and reward her from his patriarchal position without her having to point out the unfairness of her situation. Even if he did notice the injustice, he didn't do anything to rectify the situation. Ella tacitly agreed to the terms set by Philip and she accepted the high cost in order to pursue her research interests. Moreover, her boss had no compunction in expecting her to pay that price. In line with the findings of Bowles et al. (2005), Babcock and Laschever (2003) and Miller (1986), both Ella and her boss found it easier to avoid conflict and she did not try to negotiate or challenge him and she was effectively kept in her subordinate place until she made the active choice of applying for a new post.

Diana similarly met similar opposition later in her career but found a different way to cope with the obstructions. Like Ella, Diana left school at 16 with O Levels and started work in a research based public sector organization, studying and passing her professional exams. Although she had a love of science and was inspired by Alexander Fleming at school when she read his autobiography, she was not advised about developing a career:

> I just happened to be at school one day and opened a book and saw the words X Institute, and I thought oh yeah you know we're going to be moving towards that part of the world. Let's apply. So I did. (Diana)

At this institute she worked on a research project on the organism, which later formed the root of her career and where she found her niche:

> I didn't really know any other sort of [work]. I mean I knew I was interested and I loved research. I really enjoyed the research aspects more than diagnostic [work]. (Diana)

She didn't plan to do a PhD after achieving her professional exams or even know that it was possible:

> I had no idea in my own mind that I wanted to do a PhD. It just happened afterwards when I went down the research route and was doing more and more R&D and I realized yeah this is what I want to do. (Diana)

Although she enjoyed her work in the research laboratory in the early stages of her career, Diana's original boss was the archetypal medical patriarchal figurehead of the laboratory who kept the technical staff in their place. He did not see Diana as a potential scientist, brilliant or otherwise. She was a female junior technician who should know her place:

> My original boss was Dr N and I think in Dr N's eyes I was, you know, the real junior at the bench and I think had Dr N stayed on, I mean God rest his soul, I don't think I would have ever had the opportunity to do a PhD at all. He really was very much of the old school. 'Keith' [subsequent boss] was very different. He allowed you to follow your own sort of initiative. (Diana)

The new male boss encouraged her to use the work as a basis for a PhD, recognizing potential in Diana, not perhaps as creative genius but as becoming a valuable research scientist:

> My new boss at the time, 'Keith', was very supportive. The work was going well and he said to me have you considered doing a PhD so I said yes OK fine. I didn't know much about what I was taking on but he suggested I register with [university] and I did my PhD. (Diana)

Like Ella, Diana received little advice to help her progress in career terms from her family or from her school but with the support of a male advocate, she worked well and achieved status as an academic. She built up a research team, prepared grant proposals on her own projects and eventually gained an international reputation:

> It escalated from there and I realized, yes, you know I really enjoy this. I find it really stimulating and challenging. You know I really enjoy meeting international colleagues, liaising with them, and developing joint projects. (Diana)

Although Diana received help in the initial stages of her career from a male benefactor, no one saw her as a creative genius and helped to develop her further. Over time, Diana advanced her career and was able to influence on the international stage and was eventually awarded a chair in a top university but nevertheless expressed doubts about support from the head of her institute:

> I often question whether 'Paul' [Diana's medical boss at the head of her organization] actually recognizes what I do. I really do sometimes feel undervalued. I'm very well recognized and well respected externally but I just sometimes question whether he is aware of my contribution. (Diana)

In her current position, Diana considered her barriers came from her overall boss 'Paul' and the bureaucracy of the organization's processes and endorsed by him:

> I think the organization's changed. I don't like the hierarchy and I don't think they listen. I mean it's a feeling of being undervalued or something like that. I get more support from my colleagues in Europe and the US. I trust them more. I think I can open doors with my contacts and you know I think I do have a very wide repertoire of contacts, within the areas that I'm working in obviously but this place is full of bureaucratic barriers. (Diana)

Diana's bosses rejected her approaches and like Ella she was kept in her place and unable to exert the authority she wanted to. Diana was able to advance well in her career until she got to a higher level. At this point, like Ella, she was held back from making the high-level policy decisions in relation to her own subject that she wanted to make by masculinist, structural and professional barriers of demarcation that limited her decision making. She was kept in her place and unable to influence at the management level she needed to. Diana described her wish for more autonomy, which prompted a desire to leave the organization completely:

> I'd like to get out. I'd like to have more authority to make decisions. I've been asked to collaborate on a clinical programme but I need 'Paul's' approval for that and he's been very slow to show his support. (Diana)

In summary, both Diana and Ella, with the early support of male medical bosses, were able to advance to become internationally renowned scientists despite starting from the bottom rung of the ladder without a first degree or even A Levels (General Certificate of Education Advanced Level). They both found ways to counteract the effects of the hegemony in their own institutes by building networks with international colleagues. These networks gave them satisfaction and they tried to ignore the local hegemonic barriers that hindered them.

Even when they became very senior however, both women met similar masculinist and structural barriers and, like several other women, were considering moving away to escape their effects. The moves of these women were in accordance with the findings of Marshall (1995) that showed that women move to escape hostility and stress. Their working lives in their respective institutes became so unbearable that Ella eventually took early retirement. Diana survived by continuing to improve her network of contacts and international career and spent more and more time working abroad with international colleagues.

Both Ella and Diana were able to make limited transformational change in their areas of expertise, that as Maranda and Comeau (2000) propose is needed to show a level of autonomy but both felt constrained within their organizations and probably never reached their full potential.

So despite achieving success before the downward spiral at the end of their careers, Ella and Diana's rise to scientific influence took a long time and was not straightforward. Ella took a particularly circuitous route where little was offered and she had to fight and persevere to receive the recognition due to her. She had been seen as a manager of people not a brilliant scientist and expected to take on both sets of responsibilities as laboratory manager and as research scientist. Her choices were restricted by her sex and her (semi-) professional background as a biomedical scientist. For Ella in particular, after a long slow incline where she overcame many barriers, her career trajectory took a sharp downward turn and she was returned to the place of women, in a subordinate role in the medical hierarchy. Although they remained recognized nationally and internationally for their scientific contributions, their levels of influence in their workplaces declined. Both Ella and Diana were obliged to adapt to fit in with the situation they found themselves in, limited by the internal staffing and professional demarcation structures.

Arguably, the potential of Ella and Diana to become good research scientists was recognized by their initial medical bosses but this was not fully supported by later medical bosses. They did not fit the role of the masculine expert, and became the invisible 'other', each kept in her place away from the place of the decision maker and any potential for becoming recognized as a creative genius was not noticed or encouraged.

Carolyn's career trajectory was very different from the two women discussed above and she reached significant heights with several private and public sector board appointments, but in doing so she left science behind and was apparently also never recognized as being a potential scientific genius. In her mid-fifties, Carolyn was married, and like Ella and Diana did not have children. Her early school life had shown little promise and she had met many barriers including failing her 11-plus examination and an entrance examination for a private school:

> I was considered dim by my cousins who had passed the entrance exam at five years old. This has had a lasting impression on my life and career, like for example playing scrabble, 'Carolyn can't play scrabble, she's too stupid' and you know, I think I've got mental scars. (Carolyn)

Carolyn had an uneventful school life until she was 15 when a teacher spotted some potential in her:

> So I didn't do anything academically clever, but at 15 when I was sitting my O Levels I came top of the class in science and we had a woman teacher who said to me and what are you going to do Carolyn, you seem rather good at science and I said 'Oh, I don't know'. So she said 'You should think about doing a degree' and I said 'Oh a degree!' and she said 'Yes'. (Carolyn)

She did well in her A Level exams and was spurred on by the sexist comments of a male teacher:

> So anyway I progressed through 15, 16, 17, 18 and then the next telling moment in my career in school was when I was 17 in a class of boys doing A Level chemistry and the teacher, a man, said to me, 'Carolyn you should be learning how to bake cakes and sew on buttons'. So I thought, um, I don't like this very much. Anyway I applied to go to university to do science. I'd always wanted

to do medicine, but I didn't really think that I was clever enough and you can see why I didn't think I was clever enough, having been knocked on so many occasions. I went to university and got into a science BSc [Bachelor of Science] course. (Carolyn)

Carolyn continued to suffer lack of self-confidence and she refused an offer to study medicine, partly because she would have been the only woman on the programme:

In the third year the top five people were offered the chance to do medicine and I was one of them but I didn't take it. The rest were men and I felt like an outsider. Stupid really, so I didn't take it! Because again I didn't have the confidence to do it so I didn't do it. (Carolyn)

The offer came without advice or support about what accepting it would entail and Carolyn did not feel confident enough to accept. A woman lecturer, however, suggested to Carolyn that she do an Honours degree and again Carolyn was again unsure:

She said to me, 'Now Carolyn, don't you think you should do an Honours degree?' and I said 'Me? An Honours degree?', so she said 'There a lot of people with two heads who do an Honours degree Carolyn', so yes I had to think about that but she was right and really encouraged me and I got the class prize. Anyway I did the Honours degree, I did it with two boys, two guys and erm, I got a first and they didn't. I was like Hey! (Carolyn)

At the end of her first degree, she applied for posts in science institutes and someone gave her similar advice to the advice Margo received about the importance of doing a PhD:

One of the head people there who interviewed me asked me, 'Why are you not going back to do a PhD?' and I thought well I don't know and he said, 'Because you're going to find it a very limiting factor in your career if you don't'. That was another good piece of advice. (Carolyn)

So Carolyn went back to the university and the professor in her university gave his support, offering her the chance to do a full-time PhD at the same university:

Professor B who was my supervisor for my project in my Honours year encouraged me to do a PhD with him and then I did my postdoc with him and I suppose he was one of the people who was very supportive to me. He was a guy who actually encouraged me and supported me. (Carolyn)

During this time she was offered a research assistant post and then a temporary lectureship. She was also spurred on by some sexist comments from the professor:

> The professor said to me, 'I want you to see yourself in future as either baking cakes or a consultant scientist. Which is it to be, as you cannot be both?' I thought, 'I'll show you'. (Carolyn)

She then heard about a permanent lectureship in the university and decided to apply:

> A lectureship came up in the university in the department of X and I thought I'll have a go and I applied for it. I was told by the head of department in another department where I worked, well you haven't a hope in hell in getting this as it's earmarked for one of the lecturers but I went for it. I got it because they pulled another lectureship out of the bag that had been frozen. A guy actually got the first one, he was a medic and I got the other one. (Carolyn)

Carolyn was also offered the opportunity to take the RC Path exams, which set her on the path to a high-level career:

> I decided I needed to do some more clinical stuff and at the same time, the senior lecturer there said to me, you know you could do the RC Path exams and I said 'What, me?' you know the usual sort of lack of self-confidence and he said it's open to scientists, although it's a medical postgraduate qualification. (Carolyn)

Once she had passed the RC Path exams, she expected to be able to do more clinical work and approached the woman who was head of department at that time:

> Professor F said, 'Well if you think you're going to get more responsibility in here you've got another think coming, I'm not going to let you get involved in the clinical service. I want you to take on the PhD students' and I thought I don't really want to do that. (Carolyn)

Blocked by the medical hierarchical structure and demarcation barriers, a senior post came up at a nearby hospital where colleagues in a university network had heard that she wasn't happy in her lecturing post. Carolyn applied and was successful:

> Anyway I went for this post and I got it and I thought, Oh I'm giving up a lectureship, you know a prized, it's such a prized thing to be a lecturer and my husband had been a university lecturer and he'd left the university to go into industry and his mother was very upset, didn't tell the neighbours, so when I

told her I was leaving the university as well, she thought I was discarding this pinnacle of aspiration. Anyway I went for it and got it and I've never looked back. (Carolyn)

With another helping hand from a male sponsor, one of the professors in her department had moved jobs and invited Carolyn to apply for a post as his deputy in a large public sector institute and she was again successful. However, the management structure changed when this professor retired and Carolyn found herself the only woman on the senior management team and experienced being talked over (Tannen, 1992):

It was very difficult being the only woman. No one listened to what I said. Everyone talked over me and ignored me and I found it very difficult. (Carolyn)

With the support of yet another helping hand and another opportunity offered, Carolyn was then invited to apply for a post by another professor she had worked with:

'Adam' [Carolyn's boss] had said to me beforehand, you've always got a job here so I was actually applying for a post from quite a luxurious position of having got job security. So I had the opportunity to look for a new challenge, but putting your neck on the block was a big thing to do when you get further up the tree and I didn't particularly want people to know about it. I really gave it my best shot because I felt if I'm going to do it I have to do it to my satisfaction. (Carolyn)

With the support of her new boss, Carolyn continued to rise through the hierarchy, eventually becoming deputy director. She left practical research behind her but was able to influence at a higher level.

So what were the factors that helped Carolyn in her career? Carolyn's dedication to her subject was summed up by her husband's comments to her:

I just loved my job and my husband used to say, 'Don't ask Carolyn what she does, she'd work even if she didn't get paid'. (Carolyn)

Importantly, Carolyn gained useful qualifications in her Honours degree, her PhD and RC Path exams and then sought out opportunities and applied for posts. She seemed to have developed a network of local colleagues who advised her and had the support of several specific benefactors at several of these times. Carolyn was determined and persevered and developed a degree of confidence over time. She summed up why she thought she had been successful:

I think I'm quite an energetic person, I think I've got drive, energy and enthusiasm and I work terribly hard. I look back at all the things that I've achieved here and I see things that need to be done and I feel this is a new challenge and I think I'm going to enjoy being my own boss. I think, I didn't, I mean I wasn't proactive in my career. I think I reacted to opportunities. The bloke who suggested I did a PhD set me on the right path so I suppose it's getting yourself in the position that you look to the future a little bit, without actually mapping your career. (Carolyn)

Carolyn was a woman manifesting contradictions. On the Maranda and Comeau model (2000, p. 38), she seemed to exercise free will but she also demonstrated little self-confidence and was self-effacing. For instance, in the short quote above she said 'I think' five times and several times used phrases such as 'I have to pinch myself sometimes', 'Somebody's gonna find me out', 'How did I get here?' and 'I don't know how I got here'. It seemed that Carolyn did not feel she 'fitted in', in keeping with Kanter (1977) who says women consider they should be 'grateful' for what they have (p. 229). Despite reaching a high position, Carolyn seemed aware of her own difference and 'otherness'. Despite lacking a degree of self-confidence, Carolyn had high aspirations for herself, which Kanter (1977) says people require to access opportunities.

Carolyn met many barriers early on in her school life but she benefitted from several advocates who supported her over the years, some offering her new employment opportunities from which she reached a board-level appointment. With all the support she had received from various benefactors and her own determination to advance her career, had anyone considered Carolyn to have potential as a creative genius and encouraged her to develop her skills as a research scientist? Arguably not. Perhaps Carolyn's skills were always as a manager of resources and people and she was a strategist rather than a high-flying scientist who understood and enjoyed her place in that domain. Unlike Ella and Diana, she left science behind her as she rose in the hierarchy. Although the most senior woman healthcare scientist interviewed, she had not made progress as a research scientist and did not direct research programmes, unlike the senior male scientists and medics who were CEOs who retained their research activities.

DO MEN *WANT* TO SPOT WOMEN'S POTENTIAL?

In this section we show how young women scientists are positioned as operational scientists, where they are taken for granted and their bosses do not use their antennae to spot potential as a scientist, let alone a potential genius. It might be assumed that a bright young woman in healthcare

science might stand a better chance of being identified as a potential creative genius in the twentieth century and encouraged to progress. This did
not seem to be the case however, and younger women struggled to be seen
as budding scientists.

Just progressing a career was as difficult for younger women as for older
women and we now look at the incipient careers of two younger women:
Jackie and Olivia. Jackie, in her mid-twenties, thought she would be able
to plan her career, but her career progress had not gone according to plan.
She had become interested in science at school and studied for a diploma,
making rapid progress academically and gaining a First Class Honours
degree. She gained a junior post in a public sector institute and not long
after starting work had been offered the opportunity to study for a PhD by
her senior scientist boss but the offer was subtly withdrawn. Jackie realized
she would have to seek out new opportunities to find those openings that
were rarely offered even if they were not what she wanted:

> I had my whole life planned for the next five years, you know, degree first then
> PhD, that was my route and then within a year everything changed and I real-
> ized that things change so much from one year to the next that you can't even
> plan. (Jackie)

Despite having a good degree and settling in well in her research post,
any potential for developing further was not encouraged by her scientific
and medical bosses. Jackie saw her ambition for a research career fading
and the chances of developing herself diminishing as the offer to study for
a PhD was subtly withdrawn. She lost confidence in her own ability and
certainly did not see herself as a prospective creative genius and this was
far from the minds of her bosses who ignored any promise in her. As a
consequence of being in an unworkable situation, Jackie was finding her
post unbearable:

> I really do like my job and I would love to work up in the sort of work I'm doing
> but I don't feel like there's room there for me and I've actually been told on occa-
> sions that it's not my position to suggest things. They don't really want me to
> move up so I don't see how I can stay here when I feel like I'm being held back
> but I don't know where to go. (Jackie)

In this quote, as Witz (1992) and Miller (1986) describe, Jackie considered
there was lack of 'room' for her and that she was excluded from the place
of the expert scientists where her contribution was not welcomed. Despite
having shown academic ability by having a First Class Honours degree
and then achieving a Masters degree, it seemed that Jackie was kept in a
constructed subordinate place and treated by her bosses and by some of

her colleagues as 'other', as different and an outsider (see Puwar, 2004), someone who was not welcomed into the ranks of scientist and was a target of demarcationary professional barriers from her male senior scientist boss, as Witz (1992) describes. Jackie's potential was apparently not noticed and she was the victim of subtle hegemony.

Jackie's decision to leave her job seemed to her to be the only route open to her to escape what she saw as unbearable hostility:

> I had to leave to get away from the atmosphere there – it was really nasty. I liked it there in some ways but I really wanted to do a PhD and I'm the sort of person that if I have a target I set for it and I go for it. I felt really let down by 'Jim' [Jackie's boss in her previous job] when he wouldn't let me do a PhD, having said I could at first. I couldn't stand it any longer. (Jackie)

During this difficult time, Jackie found a female mentor who encouraged her not to give up her job but to take active steps to pursue her scientific career. When Jackie's career was followed up with her three years later, she had moved to another laboratory where she continued her fight to study for a PhD. She had not known whether the change of laboratory would be good or bad and although it wasn't easy she had more support and encouragement from her new boss. She was happier after the move, like several of Marshall's (1995) interviewees, who reported 'relaxation, satisfaction and continuing self-development' (p. 295). In making such a move, it could be considered that Jackie exerted a degree of autonomy in making a change (Maranda and Comeau, 2000), but it was an escape from an impossible situation (Marshall, 1995).

When followed up five years later, Jackie had gained a PhD in her new laboratory. She had been on research trips to various countries, had been overseas to set up a new laboratory, had published several scientific papers and was also leading a team of other scientists. Although Jackie had a mentor, she did not have sponsorship from a powerful advocate. Her progression was not due to being identified as a potential high-flyer but due to her own application, aspiration and persistence. She was ambitious and did not yield to the pressures of hegemonic bosses, her organization or society and her efforts may well lead to changing the organization she became a part of. She initially lacked confidence (an attribute that was so highly regarded by 'Liam, as discussed in Chapter 3) and it had not been an easy course of action for Jackie to move posts but her perseverance paid off.

Olivia was also in her twenties, a little younger than Jackie. She had had an interest in science from a young age, encouraged by her first science teacher:

You know, I wanted to please him more so I just took an interest in science and worked hard at it for years. (Olivia)

The interest in science continued:

I worked hard at school and didn't find passing exams too difficult. All my GCSEs were straight As and my A Levels were all As. When I got to my A Levels it was like Oh! You've got to choose, you've got to choose a career, a career, a career, and I didn't. The funny thing is when it came to careers evenings they don't really go on about science that much. My parents wanted me to do medicine but I wanted to study science. (Olivia)

Olivia gained a Upper Second Class degree but hadn't wanted to study for a PhD at the same university:

I felt like I didn't know anything. I haven't experienced anything. It's a really hard thing to undertake. I'm not 100 per cent sure if this was what I wanted so I just decided to get a job and get the experience first. (Olivia)

Olivia's first job was as a healthcare assistant in a local laboratory but the pay was poor. She moved away from home to work in a public sector institute where she hoped the options to progress to do a PhD related to her work might be available. She was funded by this organization to study for a Masters degree but was bored by her day-to-day work:

It's not that I feel I've mastered it but I can do it. It's ok, it's quite simple. I feel like I want more of a challenge. I keep saying I want more work. (Olivia)

Olivia was eventually given more responsibility:

They had a member of staff who has been away for a while so they gave me two of her tests which I can do in a day, so that's good. When I do things I do them accurately but I also do them quickly. I see myself as being good at time management so I fit things in, which means I finish it; if it takes someone else two days, I finish it in a day and a half. I could be doing more with my time but they are reluctant to give me more responsibility for some reason. (Olivia)

Following her Masters degree, her requests to study for a PhD never came to fruition. Her potential as a brilliant scientist may have been noticed but, if so, no one in her organization did anything about developing her further. When followed up three years later, Olivia had left the organization to study medicine. Practical research science had lost this clever, confident young woman with potential.

SOME CONCLUSIONS ON CREATIVE GENIUS IN SCIENCE

We conclude that women's potential for becoming brilliant scientists is generally not taken into consideration by the masculine elite. For women healthcare scientists, hegemony remains an active force in their lives. Overcoming these subtle and not so subtle masculinist barriers requires considerable resilience, determination and persistence. Amid the hegemony that most of the women in the study suffered, the chances of being identified as having the potential to be a brilliant scientist or a creative genius were limited, whether using examples such as Ella, Diana and Carolyn who had progressed to a high level, or the young scientists such as Jackie and Olivia, or examples from elsewhere in this book. So what needs to change for women to advance in science? How might they become lead researchers in the masculinist world of science? How can the processes that keep women in their 'place', which render invisible their potential for becoming creative geniuses, be changed?

Male advocates were vital in the part they played in the progress of three of these women discussed above but in Ella's case in particular the help was strictly under the benefactor's patriarchal control. He gave what was easy for him to give but did not push himself to give what was difficult for him to provide easily. So, individual decision makers within institutions are of crucial importance in controlling the career mobility of women. We found that even women who were prepared to fight hard and overcame many barriers to climb to the top of a career ladder were unlikely to become identified as a possible creative genius or high-flying research scientist. As shown several times in this book, several women found great difficulty in accessing part-time study or training to further their careers, hindered or stopped by (usually) male managers, Thus, obliging the women to stay in their 'place' in the lower practitioner grades and they were rarely given the opportunities offered to men in equivalent positions.

As we described in Chapter 3, while male scientists may be encouraged to undertake PhD research and given time, space and money to do this, women tend to be given support duties. So women are kept in their place, out of the research action, so as not to disrupt the masculine career course. As a consequence, women are not given opportunities because men (but not women) scientists are identified as fitting the model of gifted scientist who may become a creative genius. As a consequence, most women tend to stay in practitioner roles that include elements of caring, nurturing and support (such as 'quality' roles as Margo described in Chapter 3 and Nicola later in Chapter 6) but that are unlikely to open doors for scientists to become 'creative geniuses' or leading research scientists.

Consequently, the potential of these women scientists did not lead to them being identified as possible creative geniuses. They were labelled as 'other', outsiders who, as scientists, commanded limited respect from the masculine elite but, in the case of Ella and Diana, found support from international peers. Few opportunities were offered to them and importantly they lacked self-confidence and self-belief. With Olivia, however, her potential for development as a research scientist was unnoticed by her bosses and her confidence in her own ability led her away from science into medicine.

The main transformation that needs to happen is a massive change in culture where men are encouraged to notice that women have the potential to be brilliant. To drive this change, men as well as women need to be involved. The determination to make cultural change needs to come from the top of organizations as well as motivating staff at grassroots to believe that it is possible. Turning the rhetoric of 'diversity and equality policies' into reality is no small feat.

NOTE

1. Apart from careerism, the other masculine identities are authoritarianism (intolerance of dissent), paternalism (exercising power by cooperation and corporate identity), entrepreneurialism (accentuating competitiveness and profit) and informalism (building informal networks on basis of shared masculine interests) (Collinson and Hearn, 1994, pp. 13–15).

6.   M[o]therhood

> Many of the qualities that both men and women attribute to a feminine style
> are associated with mothering. Indeed better managers are frequently cast in the
> mould of mothers. Yet most women who *are* mothers are still absent from the
> ranks of senior management.
>
> (Wajcman, 1998, p. 70; emphasis in original)

This chapter explores how perceptions of motherhood influence the way
women are treated at work. It is not only the actuality of motherhood
that provides a barrier for women in science, but that the potential of the
maternal status is a hidden obstacle from which no woman of childbearing age can escape. Perhaps the ultimate difference that defines women is
their association with the private place of the home and their potential for
maternity, as m[o]thers or potential mothers (see Fotaki, 2011, who refers
to the term 'm(other)' and Cooper, 1992 who uses the term (m)other', each
with a slightly different spelling format, signifying the othering of mothers,
drawing on Lacanian psychoanalysis [Fotaki and Harding, 2012]). Miller
(1986) comments that from the perspective of some men, 'the home is
"women's natural place"' (p. 9). Women are often assumed to be heterosexual, expected to be caring, to raise children, to nurse the old, even to
'provide care and comfort to the tired man when he comes home at night'
(Miller, 1986, p. 75; see also Delphy and Leonard, 1992), which Haste
(1993) points out 'maintains [his] efficient functioning' (p. 67). As we noted
earlier in this book, women who fulfil caring and serving roles in science
– keeping to their place – are welcomed, regarded as non-threatening and
passive, thus allowing men to undertake the 'action' in the masculine area
of creative science where they have status as research scientists (Miller,
1986, p. 75).

How women manage the roles of mother, homemaker, housewife and
wife with a paid job is the subject of many publications over many years
(Chandler, 1991; Charles and Kerr, 1988; Delamont, 2001a; Delphy and
Leonard, 1992; Gatrell, 2005, 2008; Hochschild, 2003; Maushart, 2003;
Nutley et al., 2002; Potuchek, 1997; Martin, 1984; Smith, 1987; Williams,
2000). Nutley et al. (2002) suggest that the 'gendered nature of home life'
means lack of equality at home: a consequence of women continuing to
accept the bulk of household responsibilities is that they may not accept

senior posts in their work lives (p. 3). Liff and Ward (2001) argue that women are given the wrong messages about the promotion processes and what is required in senior posts. Many women and men identify the same issues as important but these vary in significance for them, in particular in relation to the perceived incompatibility between being a parent and taking on a senior role. Becoming a mother hinders many women from taking on leadership roles, while the opposite is often true for fathers, who prioritize breadwinning roles (Hochschild, 2003; Maushart, 2003).

Over many years, research evidence has shown that motherhood has a detrimental influence on women's careers. In our view, childcare respon-sibilities are not in themselves a cause of women's subordinate position. Rather, the predominant hegemonic masculinities within science serve to marginalize not only mothers but also those women with potential to become mothers. Such marginalization occurs due to unfair and unsubstantiated organizational assumptions that childbearing reduces a woman's commitment to her career and dulls her ability to undertake creative or groundbreaking research over a lifetime's work. In other words, women become identified as 'other' or perhaps more appropriately as we suggest here, as 'm[o]ther'. As such, in science, women are marginalized from opportunities to pursue research careers. Having the potential to be pregnant advertises a woman's femininity, which could be seen as a threat to the masculine order of the workplace (Haynes, 2008).

When women have careers and children, they conduct a juggling act of competing priorities with their careers in a way that Dorothy Smith (1987) describes as 'bifurcated consciousness' between the demands of home and work. In the following quote, Smith (1987, p. 7) is not talking about a scientific world, but her description of entering a rarefied academic life is equally applicable in science:

> The intellectual world spread out before me appeared, indeed I experienced it, as genderless. But its apparent lack of centre was indeed centred. It was structured by its gender subtext. Interests, perspectives, relevances leaked from communities of male experience into the externalized and objectified forms of discourse. Within the discourses embedded in the relations of ruling, women were the Other.

In contrast to women, the vast majority of men in science are protected by their wives or partners from the domestic care agendas, because even contemporary men remain less likely than women to be consumed by the world of home. In dual-income families, Potuchek (1997; see also Delamont, 2001b; Hochschild, 2003; Maushart, 2003) observes how women continue to take primary responsibility for household duties and care of children. Although it is acknowledged here that men may be constrained from engagement with children due to inflexibility at work (Burnett et al., 2013), it remains the case that men do less housework and mostly contribute in the more 'masculine' areas of maintenance of the home, car, and garden where the time frame for undertaking the tasks tends to have more flexibility than the home tasks undertaken by women (Edwards and Wajcman, 2005; Hochschild, 2003; Maushart, 2003). Consistent with Blair-Loy (2003), women remain under pressure to conform to the social norms of committing themselves to their children and partners rather than their careers. Arguably, such pressures help keep women in their place through prioritizing (especially maternal) responsibility for the home. In the workplace, men similarly encourage women to contribute in the subordinate caring and service image.

Women's biological differences are perhaps most obvious when a woman is pregnant. Prior to the enactment of the Sex Discrimination Act in 1975 in the UK, it was not illegal to discriminate against pregnant women and a woman could be dismissed for being pregnant. Yet, even in the twenty-first century, pregnant women still suffer discrimination at work. A recent report from the House of Commons Women and Equalities Committee on Pregnancy and Discrimination (2016–17) reported, 'Shockingly, pregnant women and mothers report more discrimination and poor treatment at work now than they did a decade ago' (p. 5). This follows research from the Equality and Human Rights Commission (EHRC, 2016a), which found little change since the 2005 report, indicating that pregnancy and maternity-related discrimination continues, despite recommendations in 2005 for improvements. In the 2016 report, over three-quarters of mothers reported that they had had a 'negative or possibly discriminatory experience' during pregnancy, maternity leave, and/or on return from maternity leave (EHRC, 2016a, p. 6). The main finding highlighted the following:

- Around one in nine mothers (11%) reported that they were either dismissed; made compulsorily redundant, where others in their workplace were not; or treated so poorly they felt they had to leave their job; if scaled up to the general population this could mean as many as 54,000 mothers a year.
- One in five mothers said they had experienced harassment or negative comments related to pregnancy or flexible working from their employer and/or colleagues; if scaled up to the general population this could mean as many as 100,000 mothers a year.
- 10% of mothers said their employer discouraged them from attending antenatal appointments; if scaled up to the general population this could mean as many as 53,000 mothers a year. (EHRC, 2016a, p. 6)

In another report from the EHRC (2016b), the cost to the state of women being forced to leave their jobs due to pregnancy and maternity-related discriminatory experiences was calculated to be between £14 million and £16.7 million mainly due to lost tax revenue and increased benefit payments. Further financial losses to the state of between £28.9 million and £34.2 million were determined in the 20 per cent of women who reported: 'financial loss as a result of: failing to gain a promotion, having their salary reduced, receiving a lower pay rise or bonus than they would otherwise have secured, not receiving non-salary benefits or having them taken away, and/or demotion' (p. 12).

For the reasons outline above, it is unsurprising that concealment of pregnancy at work is common among expectant mothers. Even the potential for motherhood may reduce their chances of employment and their status as professionals (see Davis et al., 2005; Gatrell, 2006a, 2008; Kanter, 1977; Pringle, 1998; Wajcman, 1998; Williams, 2000). Although legislation means that discrimination against women and those who are pregnant is illegal (Equality Act, 2010), unfair treatment of pregnant women at work is a longstanding and intractable issue. Much of the discrimination against women who are or may become pregnant is hidden and mostly denied by managers and human resource departments. The House of Commons Science and Technology Committee (HoCSTC) report (2014) recommended: 'diversity and equality training should be provided to all STEM [science, technology, engineering and maths] undergraduate and postgraduate students. It should also be mandatory for all members of recruitment and promotion panels and line managers' (p. 3).

We continue in this chapter by first reviewing how the state of m[o]therhood affects women's perceived place in the workplace. We then look at how our interviewees experienced m[o]therhood and how they coped with discrimination in the workplace due to their maternal status, and how they maintained a career with pregnancy and children, including the need to maintain publication rates. We also look at two somewhat different situations: the first is the experience of a male carer who suffered similar discrimination to that of the women but considered he was treated worse; the second situation is examining how well lone mothers coped with balancing their careers with children.

M[O]THER

We have already observed how, while masculinity in science is defined as 'positive', women are, through their female bodies, defined in the 'negative', as 'other'. As a result of being assigned a lower social status than men, women are deemed to be of less importance than men, and women's health, social and intellectual needs are as a result habitually ignored (Annandale and Clark, 1996). Davies and Thomas (2002, p. 479) argue that for many women 'conceptions of being positioned as "the other", of not fitting in to the "masculinist ideal", [are] very prominent. For many women, showing feminine attributes is something that always needs to be managed, as it can be seen to silence their authority and their involvement in decision making'. In line with Davies and Thomas (2002), we have thus far argued that a masculinist discourse exists in science where women are assigned a lower status in comparison with male colleagues and are 'positioned as "Other", because they do not "fit into . . . the masculinist ideal"' (ibid.). As a result of the lower status ascribed to them, women in science are expected to fulfil subordinate and inferior roles to men.

As in many organizational contexts, one form of women's 'otherness' in science relates to actual, impending or even potential motherhood. The HoCSTC report (2014) included evidence from universities about the detrimental effect that maternity leave has on women's science careers largely because of lack of organizational commitment to introducing a fairer system, for instance in providing cover for women when they are away, lack of encouragement for shared parental leave and lack of provision for integrating women back into the workplace after maternity breaks.

Some ten years ago, in 2007, the UK government produced *Fairness and Freedom: The Final Report of the Equalities Review*, which sought to understand why women fail consistently to achieve the same occupational advancement as men within organizations, even if they have equivalent skills and qualifications. This report concluded: '[T]here is one factor that above all leads to women's inequality in the labour market – becoming mothers' (Equalities Review, 2007, p. 66).

It is acknowledged here that the *Fairness and Freedom* report encompassed women workers across a range of occupational statues, organizations and fields – it was not concerned only with women in science. Nevertheless, the above observation accords with evidence from the literature that motherhood is used, within organizations, as justification for women's continued limited career progress compared with equivalent males. Acker (1990), Ashcraft (1999), Bailyn, 2004, Budig and England (2001) and Haynes (2008) substantiate this argument. Once they become mothers, women are consistently disadvantaged within labour markets, earning less than men and receiving limited career advancement compared with male colleagues (Blau et al., 2014).

Research shows how negative organizational attitudes about women with children persistently adversely affect mothers, because employers unfairly correlate pregnancy and mothering with disrupted workplace routines, reduced commitment to paid work and intellectual instability. Such unsubstantiated beliefs are evidenced in studies about organizational undervaluation of mothers' abilities and their work orientation. Thus, it can be shown that recruitment panels commonly privilege male applicants (Adler, 1993; Cunningham and Macan, 2007; Kanter, 1977; Schein and Davidson, 1993), how organizations (sometimes subtly, sometimes overtly) may downgrade the role and position of pregnant women and new mothers (Ashcraft, 1999), and how colleagues may often assume lowered workplace commitment among employed mothers, even when they have no evidence to prove this (Haynes, 2008).

Women's capacity for reproduction is assumed within many organizations to compromise their intellectual focus (Annandale and Clark, 1996); conversely, men's bodies are disassociated with the labour of infant reproduction and care, and men, as a consequence, are treated as intellectually sharper and more measured (Witz, 2000). Women's 'maternal' bodies are associated with reproduction and the care of dependent children (sometimes irrespective of whether they are mothers – women's potential for childbearing is sufficient for employers to regard them as maternal bodies). Maternal bodies are regarded as 'intrusive', a disruptive influence on organizational practices (Acker, 1990, p. 152; see also Tyler, 2000; Warren and Brewis, 2004).

As we demonstrate in this chapter, the pregnant and newly maternal body may also be treated as especially unwelcome in workplace settings. Maternal bodies that bring into the workplace reminders of the dependent bodies of infant children are treated as 'other' and colleagues may feel fearful, disgusted or both by the prospect of what feminist sociocultural literatures refer to as mothers' 'leaky' bodies – prone to embodied and hormonal changes, as well increased liquidity in the form of amniotic fluids, breast milk and, potentially, vomit and tears. The maternal body, especially pre- and post-birth may thus be seen as 'taboo' at work to the point where it becomes a source, if not a focus, or organizational abjection (or disgust), leading to workplace shunning of pregnant and newly maternal women in professional settings (Fotaki, 2013; Tyler, 2000). As Höpfl (2000) has observed, 'the organization [is] not a place for women with physical bodies which produce . . . breast milk and maternal smells' (p. 101). Several women spoke of being ignored, or 'wallpaper' when pregnant, as Alice, a senior scientist who could be described as 'in charge' recalled:

> I can remember being at work and pregnant, heavily pregnant when some bloke walked past me saying could they speak to the person in charge, you know as though I was a piece of wallpaper. (Alice)

Similarly, Angela commented:

> 'Sam Cook' introduced him [professional contact] to 'Richard' and they had a chat. Then he [professional contact] came past my office, which was next

to Richard's – I was sharing it with 'Carrie' [secretary] and he assumed I was
Richard's secretary. It was just because I was a woman, not helped by the fact
that I was pregnant at the time and that sort of thing really annoys me. (Angela)

In contrast with such experiences, men's bodies appear to be valorized within organizations for their 'minimal responsibility in procreation'
(Acker, 1990, p. 152). The masculine body is associated with balance and
stability. Rationality and intellect are constructed as masculine characteristics and in dominant 'rational' cultures (of which science is a leading
example) masculinity dominates.

SHALL I, SHAN'T I?

The literature suggests that women are jeopardized in their careers
because they have or may have children in the future (Davis et al., 2005;
Gatrell, 2006a, 2008; Pringle, 1998; Wajcman, 1998; Williams, 2000). It is
thus understandable that 'the majority of women managers maintained
that positively resisting the imposition of the mother role at work was a
necessary tactic, if one wanted to be treated seriously as a "manager"'
(Davidson and Cooper, 1992, p. 91).

Above we have noted the literature that shows that women applying for
jobs may suffer discrimination on the basis that they may become pregnant, even if they are not mothers at the time of their interviews. These
women who are not mothers may also be treated as inferior within labour
markets due to their potential for childbearing in the future. Reluctance
to recruit potential mothers made national news: a survey undertaken
by the law firm Slater & Gordon reported that 40 per cent of managers
avoid hiring women of childbearing age, which could be taken to mean
from age 16 to age 50 (*The Guardian*, 2014). This was also highlighted in
the evidence accepted in the HoCSTC report (2014), where the British
Pharmacological Society highlights that 'many women in STEM suffer
bias due to expectation, in that the potential for a woman to take maternity
leave or to require flexible working in future can impact the judgement of
interviewers' (p. 19).

Paradoxically, women who do not have children (whether through preference or circumstance) may be regarded as different or odd by colleagues,
since they are seen to be failing social expectations of what women ought
to prioritize in life, primarily the bearing and raising of children (Davidson
and Cooper, 1992; Ramsay and Letherby, 2006).

So perhaps unsurprisingly, it is worth recording that some women
rejected motherhood. Two women, Jackie and Sarah, had decided against

having children because they wanted to concentrate on their science careers, though in Jackie's case, she still suffered discrimination.

Jackie, in her twenties and at the start of her career, and Sarah in her thirties were concerned at the prospect of giving up their careers to have children and had made conscious decisions against having children. Jackie mused on the consequences that child-rearing would have on her career aspirations:

> [B]ecause you know, I put so much of my life towards my career and fighting for things, you know, would I really want to give that up? Well it's not a case of giving it up but you do have to compromise and you have to sacrifice to make to make your life balance. (Jackie)

Significantly, Jackie saw 'fighting' as an integral part of her working life; she didn't want to give up work and contemplated the need to 'sacrifice' some aspects of her personal life for her career and that included the decision not to have a child. Both Jackie and Sarah were in long-term heterosexual partnerships and in Sarah's case, her partner already had a son from a previous relationship, which Sarah regarded as 'quite enough responsibility thank you very much'. In Chapter 3, we described the discrimination Jackie suffered, ascribing it to subtle masculinities enacted by senior men over women's careers. However, it is also a possibility that her male scientist boss regarded her as lacking commitment because of her potential for motherhood.

Like Jackie and Sarah, several women scientists struggled with the notion that they might have to give up their career if they had children but not to the point of deciding against having children. In Lisa's case, before she became pregnant she identified that her career was so important to her that she and her husband thought seriously about whether or not she should have a child:

> Scientists have a terrible fear that if you stand still for a moment you're going to be left behind but I took the risk and had a baby. (Lisa)

Lisa, now a professor with a young daughter, was highly concerned that having a child would affect her career:

> I got married when I was 33. We sat down and said should we have a child or not. I was absolutely petrified, thinking what's going to happen to my career? (Lisa)

Significantly, while there has been a marked increase in research on paternity and work–life balance, much of this focuses on barriers faced by men who seek to manage their traditional breadwinner role alongside that of involved parent. While some recent research demonstrates paternal concern about maintaining career advancement if working part-time (Gatrell et al., 2014), it would be unusual to find evidence of men deciding against parenthood in case it may have a detrimental effect on their career, as Lisa disclosed.

Of the 13 women who took less than or no more than the statutory maternity leave, Lisa's situation was notable in that she worked on the day her child was born and went back to work (unofficially) two days later:

> The morning I gave a lecture and in the evening she was born. Two days later I was back in the lecture theatre because my students were in their final year. In some ways it was because my husband was open and flexible that it meant that I did not have to feel guilty about my child but I couldn't let down the 60 students I was teaching. (Lisa)

Lisa highlighted the problem many women find if they are highly committed to science and don't want to lose the impetus of the scientific career progress when they have children. Her solution was to take off only a few days and to go back to work without official approval because she didn't want to let down her students. She may also have felt concerned about loss of her role and responsibilities should she take significant time away from work. As Millward's (2006) study shows, in circumstances where maternity cover is provided, this might involve a 'permanent re-shuffling' of (often the most interesting) responsibilities among colleagues to 'ensure adequate cover' (Millward, 2006, p. 323). Thus, women returning from maternity leave might find themselves expected to continue with the more unpalatable aspects of their jobs, this making them feel undervalued and 'deskilled' while the more cherished and appealing tasks have been reallocated to others in their absence (see also Haynes, 2008).

In accordance with Miller (1986) who notes that women and other subordinate groups prefer to keep a low profile during their pregnancy to avoid conflict, Lisa did not ask her line managers about the maternity entitlements in the light of her pregnancy and imminent birth, and they didn't offer her the information. Experiences such as Lisa's were highlighted in the evidence to the HoCSTC report from the Institute of Physics, which states that there is 'anecdotal evidence from many of our members in academia that maternity leave is often organised ad-hoc, poorly implemented at the departmental level and women are not properly informed of their entitlements' (p. 36). Lisa just proceeded the way she felt was right for her; she didn't try to negotiate. She might well be justified in being sceptical of support from her managers.

Other women viewed having a child as something that they would enjoy alongside their career but recognized that there would be sacrifices in both their careers and in their child-rearing. From our general knowledge as well as from our research, the working hours and commitment of scientists has hardly been affected by legislation stating that people should not be working extended hours: both female and male scientists do not seem to have changed their working practices and continue to work long hours. For some young women, the demand to work long hours in science worried them in anticipation of having a family, as Olivia confided:

> I think a lot of the time when I stay really late or take work home, that's something I would have to change somehow if I had a baby. (Olivia)

It was not that the women with children did not like being mothers. Elena, in her early thirties with a PhD and a baby daughter, and Carla, a single mother in her forties with a son of eight, both used the words, 'I love being a mum'. Angela had six years away from science looking after her three children and for her, having children improved her confidence and changed her life:

> Having children I think changed me tremendously. It was the most challenging thing in my life. I only recognized it when I went back to work afterwards. I wasn't the same sort of person as I'd been before and I became more assertive, for instance. (Angela)

Essie, in her fifties, a scientist turned corporate manager and Angela, a senior scientist in her fifties who had achieved a PhD in her forties, went further. The emotions involved in having a baby are usually dramatic (Haynes, 2006) and Essie and Angela were two of several women who described a transformation in themselves when they had babies. Essie, who had four sons, and explained how nothing had been as important in her life as having children:

> Each time that I've had a child, having a child was the most important thing in my life. Once the baby became a child, I could concentrate on my career again but then I went on to have another three. (Essie)

It seemed that once Essie children had passed the baby phase, her career again became very important in between having each of her four boys.

We move on now to describe how some of our interviewees were discriminated against due to their maternal status.

COPING WITH PREGNANCY DISCRIMINATION

Unfair treatment of pregnant workers, in particular, is described by Mäkelä (2005) as 'commonplace' (p. 50) and, in the UK, we have already noted that as many as one in nine (11 per cent) of mothers or pregnant women a year received 'harassment or negative' comments related to their pregnancy (EHRC, 2016a). This occurs for a variety of reasons. It is well known that any gendered 'discounting of women's abilities' is likely to be 'exaggerated when women are pregnant' (Halpert et al., 1993, p. 650). This is because co-workers tend to imagine that women's skills, intellect and work focus will be dimmed by pregnancy. As a result, pregnant women, both in science and more generally, may be excluded from interesting and career-enhancing projects due to colleagues' unsubstantiated fears that their inclusion might downgrade overall group performance (Gueutal and Taylor, 1991). It is worth noting here that, in the study by Halpert et al. (1993), colleagues' assessment of women colleagues capabilities 'plummeted' if they became pregnant, yet the research (in keeping with our findings) demonstrated no evidence of the truth of such judgements, which were shown to be unfounded (p. 650).

For all women there is the 'biological fact [that] childbearing introduces an immediate disjuncture in the notion of career progression for women' (Höpfl and Hornby Atkinson, 2000, p. 135; see also Evetts, 2000; Hochschild, 2003). The mothers in our study had varied experiences stemming from the length of maternity career breaks they took and whether they returned to working full- or part-time. There were also common experiences of lack of support from line managers. Such omissions were often subtle in how they were manifested and the women accepted lack of support as the norm rather than framing this as discrimination.

Of the 17 women with children who participated in our study, most (13) took between a few weeks and the maximum of their paid and unpaid entitlement. Four others (Angela, Alice, Elena and Mary) stopped work for longer than their statutory maternity leave. Six women (Mary, Nicola, Rosa, Elena, Gina and Jane) returned to work part-time. We now look at the experiences of some of these women, first discussing how the very fact of their pregnancy affected their lives at work.

Although pregnancy and family dominated the lives of women for a period of time, the women scientists found it difficult to forget their work obligations and experienced continuing demands from their scientific lives. During this time, they also experienced discrimination at their places of work that they did not always recognize as such. Nina, in her early thirties, said she believed that the discrimination she experienced was not because of her gender but due to a lack of career structure and training programme for research healthcare scientists in her public sector organization:

> I know of a Caucasian white male who is in a similar position to me who is effectively stuck now at the same level and he's not been offered any form of career advancement or training. (Nina)

Despite Nina suggesting her male colleague suffered the same barriers as she did, as described below, Nina's interview was peppered with experiences of discrimination on account of her pregnancy, in accordance with the research by Davis et al. (2005), Gatrell (2008), Pringle (1998), Wajcman (1998) and Williams (2000), as well as several government reports (including EHRC, 2016a; Equalities Review, 2007; HoCSTC, 2014) that highlight the likelihood that pregnant women will experience discrimination. This discrimination damages women's careers.

These experiences among women in science are in keeping with wider research, which consistently demonstrates the disparity between equal opportunities policies (aimed at seeking to prevent discrimination against mothers) and the treatment experienced in practice by employed pregnant women and mothers (especially those with infant children). Pregnancy and new maternity mark the point at which women are most likely to face barriers to career advancement, even when working for organizations with supposedly family-friendly policies in place (Blair-Loy, 2003; Longhurst, 2001, 2008, James, 2007; Rouse and Sappleton, 2009).

Davis et al. (2005), Gatrell (2006a, 2008), Pringle (1998), Wajcman (1998) and Williams (2000) all describe how pregnant women suffer discrimination. In 1977, Kanter indicates that women will also conceal their pregnancies to become invisible. Perhaps unsurprisingly, contemporary women professionals, including scientists, continue to pursue strategies of secrecy and silence (Gatrell, 2011a), delaying the announcement of pregnancy and concealing ill health in attempts to avoid being unfairly marked out as either less reliable, or less competent than colleagues (Gatrell, 2011b; Longhurst, 2001; Mullin, 2005; Warren and Brewis, 2004).

Nina, as 'm[o]ther', for instance, said she was normally outspoken on such matters but preferred to keep a low profile in the early stages of her second pregnancy rather than risk a clash with her boss (Miller, 1986). She may well also have been hiding her pregnancy as it highlighted her body as that of a sexual being (see Martin, 1990), a situation that was out of place in the science work environment. By keeping quiet, however, all she did was delay the trauma:

> When I got pregnant the second time I thought it best to keep quiet and keep my head down. I didn't tell anyone I was pregnant till I was about six months and

began to show. We [Nina's husband and Nina] thought it best to have the two children close together so that I could get back to concentrating on my career but I know they weren't happy about it. I wore baggy clothes but no maternity dresses until I couldn't disguise it anymore. (Nina)

Discrimination against Nina extended to being denied training: she had recently been told by her boss that there wasn't any point giving her any extra training while she was pregnant, which seemed to her to mean that her career could not progress during her pregnancy:

I was basically told that all my training was on hold because I was pregnant, which as far as this organization is concerned basically means my future is on hold. (Nina)

Nina's boss had also told her not to be so career focused:

He said I should slow down a bit and stay at home and look after my babies and I shouldn't be so career focused as I'm still young and have got plenty of time. (Nina)

Nina faced the dilemma of dealing with a sick child when she was doing her research projects:

Then my child was very young he got quite poorly with picking up everything at nursery and I had to have a lot of time off to look after him. Normally I would have one sick day a year, if that, whereas suddenly I have had to have an awful lot of time off, which I find hard because I feel guilty for not being there and I feel guilty for not being with him if he's poorly. I take some of it as carer's leave and some as annual leave but that doesn't lessen the fact that I'm not here when I've got a project running. I find it very difficult asking someone else to pick up something when I'm not here and they've got their own work to do. (Nina)

In keeping with the assessment by Gatrell (2005), who found that some women move to new jobs based on the new situation of having children rather than challenging their existing employer, Nina was so upset by the attitude of her bosses and organization, that she was considering leaving her post, giving up science altogether and becoming a teacher, as two of her colleagues had recently done:

So, being brutally honest, I'm at the stage now where I'm thinking of completely giving up working for [organization] and completely giving up being a scientist and trying to think of something I can do which is family friendly. I know two scientists who have left here in the last few months to become science teachers. (Nina)

Nina's position was becoming intolerable and she could not face the continued pressure:

> I think I'm just a bit disillusioned at the moment. I've given up. I am not sleeping at the moment because I'm worrying about the work that I won't be able to do before I go off on maternity leave. (Nina)

When Nina returned to the laboratory after her previous maternity leave, she had found that no one had undertaken her work while she had away, providing her with a backlog on her return. About to go on maternity leave a second time, Nina's stress levels were high for several reasons, including finding someone to undertake her work when she was away the second time:

> I'm worrying about the girl who has agreed to take on my job in addition to her current job while I'm not here and how she will cope. (Nina)

This quote is also a good indication of the commitment women with children feel towards their scientific work: Nina still felt an obligation to make sure the work she did was carried out satisfactorily in her absence but did not want to draw attention to herself (Miller, 1986) by complaining about the work that had been left. In addition to the pressures on Nina, there was the added pressure for the woman who was covering for her, taking on Nina's job in addition to her own role in Nina's absence. Such issues are highlighted in the HoCSTC report (2014), noting evidence from several universities showing how maternity leave was poorly organized, work wasn't covered during maternity leave and that women were expected to catch up with work on returning to work. Significantly, it was a woman who had 'agreed' to take on the extra work in Nina's absence; there was no automatic replacement to cover maternity leave in Nina's organization, provoking problems for those leaving and those providing the cover. Under such circumstances, some new mothers compromise on their maternity leave, working throughout their maternity entitlement either from home or by shortening their leave and either working late into pregnancy or returning to work early so as not to place too much burden on colleagues, or to leave work undone (Gatrell, 2008).

In keeping with Nina's experience, respondents in Millward's (2006) research on employed mothers and maternity leave showed how mothers felt guilty about the inconveniences caused by pregnancy in terms of placing additional burdens on colleagues. Like Nina, Millward's respondents recounted feeling anxious and pressured due to fears that they may be failing their employers and adding to colleagues' workload. In Nina's case, interestingly, she saw the arrangement of maternity cover as her responsibility, not that of her line manager. She also seemed to see the responsibility of organizing a job share as down to individuals, commenting:

> I'm aware of two women, one in her forties and one who has just had a baby in her thirties who would both like to work part-time but I suppose they've not spoken to each other about it because that would be an ideal opportunity to job share but neither of them feels able to mention this to the director. (Nina)

Unlike in academia, Nina's organization did not have a career path or training plan for the research healthcare scientists. Nor did they have any arrangements whereby scientists on maternity leave could keep in touch or be retrained at the end of their maternity leave. In Nina's organization, women who came back to work after maternity leave were expected to pick up where they had left off.

Consistent with Gatrell (2005), who reports that women's motivation is not diminished by having the responsibility of children, Nina did not want to 'slow down', she wanted the opportunity to progress her career but she also wanted to have a family. Because of the difficulties she was facing, Nina felt forced to consider a complete change of career into teaching.

The pressure on women came not only from work but financial matters were also foregrounded. To place a second child into a nursery as Nina had done with her first baby, plus travelling expenses, would leave her with a sum of money that made her feel that it was not financially viable to continue working. The way Nina and her husband viewed the family accounts, where responsibility for paying for childcare costs were seen to come from Nina's salary, fits with the Potuchek's (1997) observations that women rationalize allocation of the wife's salary as a 'gender boundary that distinguishes their husband's employment from their own' (p. 63). However, by putting in place such boundaries, the wife's financial contribution is belittled. Interestingly, according to the published national statistics, Nina's salary alone was approaching twice the national median for full-time employees (women and men combined, according to the ONS, 2016), which bodes ill for low-paid women who have children and who want (or need) to work.

RETURNING TO WORK PART-TIME

It might be supposed that post-birth, the workplace treatment of new mothers would be less discriminatory than that of pregnant workers. However, research across the arenas of management and policy indicates that women returning to work from maternity leave continue to experience discrimination, especially if they switch from full- to part-time contracts. Particularly among mothers working part-time, marginalization of women's careers and retrograde revision of their workplace status is commonplace as they are parked on the 'mummy track' where the prospect of career advancement is limited (Ashcraft, 1999; Blair Loy, 2003; Gatrell, 2007; Haynes, 2008; WWC, 2006b). Halpert et al. (1993) report how one women manager who became a mother and who had previously been described as a 'superb' colleague was, post-maternity leave, regarded by her line manager as a 'terrible' colleague.

We now look at the varied experiences of the six women who returned to work part-time after their maternity leave (Mary, Nicola, Rosa, Elena, Gina and Jane), the first three of whom suffered because their part-time work was apparently a contributory factor in being regarded as lacking commitment to their roles.

The female head of the whole department was initially very supportive of Mary's return to work on a part-time basis as a biomedical scientist post after a substantial break from laboratory work of six years. Later Mary wanted full-time employment because she realized that a permanent full-time post would help her be recognized as a scientist. Although she had succeeded in having the post made permanent, she was unable to increase her hours to a full-time contract:

> 'Penelope' [department boss] invited me in to talk to her and she said, well you're just the sort of person we want. She basically said, 'What hours do you want to do?' So I told her that initially I wanted to work only three days a week from 9:15 until 3 o'clock. That was good and it eventually became permanent but I couldn't get them to make it full-time. (Mary)

Interestingly, Mary here expresses gratitude to her female boss of the department for making the part-time post available for her, but later, when the same female boss refused to make the post full-time, Mary declined to suggest it was the female boss's responsibility but blamed 'them', implying the more ill-defined organization. After several years, her post became permanent but not full-time and as she overcame one barrier another was erected. Although she worked 30 hours a week (instead of the usual 37), it seemed that her male medical boss saw her part-time employment as a yet another sign of her mothering responsibilities, which contributed to her unacceptability for studying for a PhD.

Mary's experiences of discrimination were different from Nina's but also involved denial of access to her further development. In her early fifties, Mary was a late starter in science with teenage children, a biomedical scientist working in a research post and undertaking an equivalent role and similar responsibilities to other designated research healthcare scientists. This was not an issue for anyone until Mary decided it was time for her to advance her career. Mary had demonstrated ability both academically and in her work role, which had gained encouragement from her male medical boss who initially offered her the opportunity to study for a PhD. Although he recognized in principle that she would be capable, she was having great difficulty persuading him that she was serious in wanting to study for a PhD. Demonstrating subtle masculinities, professional discrimination and sexism in practice, he subsequently suggested it would be better for her to stay as a biomedical scientist. In addition, seemingly being considerate of the welfare of Mary and her family, her boss advised her that she should concentrate on her family commitments and should not be thinking about starting a PhD. His gendered comments about her responsibilities to her family contravened equality and diversity legislation but that did not seem to matter to her boss. Citing motherhood as a barrier to Mary establishing a research career, he barred her from undertaking a PhD, which would have opened the door for her to be recognized as a research healthcare scientist:

> I've got several ideas for a PhD but they are trying to keep me down. 'Derek' (Mary's boss) says I can't do it because I'm a biomedical scientist and he says it would be better for me to give my attention to my commitments with the family. (Mary)

Mary had completed a Masters with distinction in her late forties:

> When I did my MSc [Master of Science] I said I would do it in three years and I did it although I could have gone on for another two years. I don't mess around. When I say I'm going to do something, I do it and I knew I was going to get a distinction and I did. I want to do a PhD and he [her medical boss] said I can but it's always 'We'll talk about it in the future'. (Mary)

Mary gave her boss several reasons why doing a PhD would help both her and the organization. One of these reasons was that Mary realized she would not be taken seriously as a scientist until she had a PhD:

> I really don't think he understands my situation, I mean he said, 'Why do you want to do a PhD, it's not going to give you a better salary you know. Post-docs are not any better paid than you are' and I said, 'Well, it's not that; it's because for one thing I don't really think you're taken seriously as a scientist unless you have "Doctor" in front of your name'. (Mary)

Although promotion was not her immediate aim, Mary also realized that she could only lead a team of other scientists if she had a PhD. The importance of this became clear when she wasn't credited as first author on a paper on a R&D project that she had led and where she had written the first draft. That women are less likely to be first authors is one of the manifestations of unconscious bias cited in the supplementary evidence from the Open University to the HoCSTC report (2014). Mary felt she was treated as a support worker:

> I've worked on projects where I've done most of the work, written a paper and the clinical scientist who was involved in the project put his name as first author on the paper and he gets all the credit for the work that I've done. (Mary)

Because of this persistent and subtle discrimination that he directly related to her responsibilities for her children, she was considering changing jobs away from R&D into clinical diagnostic work where the likelihood of studying for a PhD would be much reduced. She tried challenging his decision over a two-year period, gaining support from other senior colleagues including her boss's boss, but no one was effective in persuading this medical doctor to give her the positive support she needed so that she could register for a PhD.

Mary worked in an R&D laboratory and at one level was highly regarded in that she delivered innovative work of a high standard. However, her request to study for a PhD was blocked by her boss despite his initial encouragement and she was isolated overtly because of what her medical boss regarded as her maternal responsibilities. Mary was also being discriminated against by her boss as her professional background in the constructed gendered role as a biomedical scientist also acted to preclude her: the barriers of professional demarcation (Witz, 1992) were a second barrier excluding Mary from progressing as a scientist. Mary said she was exhausted. She felt aggrieved that her boss and her organization were ignoring her academic achievements and her work record:

> It gets to the point where they just wear me down. They take and take and don't give anything back. (Mary)

So her male medical boss initially offered Mary the chance to do a PhD but subsequently changed his mind; perhaps he was nervous about a part-timer achieving the status of scientist and wanted to keep her in her subordinate place as a biomedical scientist. How to confront him was an issue. To challenge her boss by using a formal route using equality and diversity legislation was not something Mary would consider to demonstrate her grievance saying, 'that is out of the question' (see Gatrell, 2008). She 'knew' that 'taking out a grievance' through the organization's internal procedures would only alienate him further. Mary was disillusioned with her boss and her organization and did not know whether to continue the 'fight', asking, 'Do I really want to go on fighting the battle?'.

Mary eventually had enough of this unsympathetic doctor and, in a similar way to other women in the study, was applying for other jobs to get away from her indifferent and discriminatory male boss. The moves of these women were in accord with the findings of Marshall (1995) who found women left posts because of hostility, isolation and stress, although Marshall did not identify lack of access to education and training as a primary cause of the antagonism as Mary and others had encountered.

Mary's relationship with her boss had broken down and no one gained. Initially considering moving to a post in a clinical diagnostic laboratory where the work would not lend itself to a research degree, when followed up three years later, Mary had changed direction in her career to work in a scientific writing role and so was lost completely to practical science and the organization. Due to her maternal status, she had been pushed out from what her medical boss considered to be a high-status masculine role (Miller, 1986, 2011). She had tried to be an active chooser but lost out to the subtle hegemonic authority of one individual in the powerful medical profession. Although he was only one individual, the careers of many of the women in our study were hindered by other similar individuals who exerted professional and structural barriers of demarcation regarding how suitable or otherwise it was for mothers to pursue a scientific research career (Witz, 1992) acting within the masculinist power of a hierarchy that places medical doctors at the top of the professional tree.

Nicola, in her forties, had two children in her early thirties and worked part-time for four years after the birth of her first child. When she returned to work, she moved to a new laboratory but was obliged to drop a grade because she was working part-time. Her female medical boss did not accept that Nicola could have a management post while she was working part-time and suggested she apply for another more highly graded full-time post later. Nicola said working part-time and earning less 'suited me at the time' and considered that she had little choice at that time and had little chance to exercise autonomy. She was pleased to start work again and like other women did not have a negotiating position (Babcock and Laschever, 2003). While working part-time, she was then asked by a male medical doctor to cover a more senior colleague's maternity leave:

> 'Janet' went on maternity leave and I was asked to cover her job; that was good for me but I didn't get paid for acting up. (Nicola)

Although Nicola accepted that it was to her advantage to 'act up' in a higher-level post so that she would get more experience, and despite being not recompensed for taking on the more senior role, she took on the challenge. Nicola subsequently took a science Masters degree and then a management Masters degree. She did not work in a research laboratory at that time so working for a PhD was not considered and at that point had found it difficult to find a new post or to gain promotion above a middle laboratory manager level. Her solution after two years was to move to a post in another laboratory with responsibilities for quality that gave her a slightly higher grade (although not the grade she had sought) but which moved her away from science, in much the same way that Margo, Essie and Mary moved away from science. Some years later, when her children had left home for university, Nicola gained a senior management post in another hospital, successful by her persistence and willingness to move away from her family to return home at weekends. Arguably, Nicola's career was disadvantaged by the subtle discrimination she experienced when returning to work part-time.

Rosa took a different approach by returning to work part-time. Although she took only a few months maternity leave she took a job share after the birth of her second child:

> I didn't have a career break as such but I did do a job share. I came back and worked full-time after my first child but when I had my second, a colleague 'Claire' was looking to share so I thought I'd just do it for a few years and it would be good for me and good for the family to try and balance everything out but I'd never got any intentions of it being a permanent thing. It was like putting my career on hold temporarily so I did that for about six years but it was difficult to stop without letting Claire down. (Rosa)

Although it seemed a good idea at the time, it was difficult later for Rosa to return to full-time working because there wasn't a post available without letting down the woman with whom she shared the job. As with Nicola and Mary, it also indicated a lack of support from her bosses who might have employed another part-time person to fill Rosa's slot, rather than expecting Rosa to take responsibility for the job share. Rosa was in a similar position to Nicola and Mary without a PhD and she found it impossible to change direction from diagnostic pathology to pursue a career in research science. Because she had wanted to stay in science, she had not gained a management qualification that might have helped her gain promotion as a laboratory manager, but that would have meant her focus was on management rather than the science she loved and wanted to pursue but was subtly blocked from doing so (more of Rosa's story is told in Chapter 3).

Elena, like Lisa discussed earlier in this chapter, saw her maternity leave as asking 'favours', not an entitlement based on being a working woman in the UK. Although being clear that her commitment to her career was critical to her life, emphasizing the particular difficulties working in science brings to a career where experiments cannot be left. While Elena was on maternity leave, she took steps to keep up with the science so she never really stopped working, although this was difficult:

> I was getting the journals to read and the occasional paper if someone visited when I was off. We were writing up a chapter for a book so I sent that for reviewing but I can't say that I was running to the library to keep up with the journals. I just physically couldn't do that. (Elena)

Elena was worried what would happen if her baby was unwell:

> Leaving her was quite traumatic for me though probably not for her, just to make the decision was quite stressful because you are with your baby 24 hours and suddenly you have to make a choice of arranging childcare, which is not the easiest thing, and then what if she's not well? What if the child minder is not well? You worry so much before you actually start the process. (Elena)

Like Nina, Elena expressed concern about organizing her scientific work, particularly the practical aspects, around her baby:

> Basically I had to rearrange my lab work because of the pregnancy because certain chemicals were a potential problem and so I had health and safety issues and all of that to worry about. (Elena)

Like the other women, Elena wanted to maintain a low profile and working part-time was an issue:

> It was also a problem for me working part-time as I had to manage doing experiments within a shorter time frame. I didn't like to ask people from the lab to help me but they were extremely helpful just picking up the experiment from the moment I couldn't carry on but I didn't want to ask any favours. (Elena)

Assisting women on, and returning from, maternity leave was not something Elena's organization accepted as their responsibility. Helping women who were working part-time was apparently regarded as being the responsibility of the individual to manage. Although their lack of supportive actions were not discriminatory, action to assist women back into science would have been beneficial to women like Elena and would also assist the organization. Again, like many women scientists, Elena was concerned about how she was going to manage her scientific career with a child:

> Basically I worried about how my work will fit in with the childcare arrangements and to start with I was thinking more like going five days a week but on

fewer hours but from the point of the techniques I have to use in the lab so it was really not feasible because you can't stop your experiment because you have to go and pick up your baby so I work three whole days but it was difficult changing to working part-time. (Elena)

Like the other women scientists with children interviewed, Elena wanted, and expected to have, the best of both worlds: she wanted children and wanted to pursue her career in science. Significantly, women are not alone in this regard – as noted earlier, men in science and technology are increasingly likely to seek quality relationships with dependent children, especially in the early years (Burnett et al., 2013). Unfortunately, the public sector institutes in which most of our participants worked (with possibly the exception of Gina's organization, as described below) did little to create a culture where parents with children were supported. Help with additional training on returning to work after maternity leave was not available nor did the organizations seem to believe that they had a role in supporting staff with maternity leave and child care arrangements.

Two women, Gina and Jane, were content with long-term, part-time working: Gina chose to continue to work part-time long after returning from maternity leave, and Jane, a medical doctor, worked part-time for several years after the birth of her second child and then changed to working full-time without difficulty.

Gina studied for her PhD full-time before starting work in her public sector organization and had reached a high level with a national and international reputation as a senior scientist and head of a large department. Nevertheless, she was only one of four women scientists (of a total of 15) at her level in her organization and there were no women directors at the level above.

Gina chose to work part-time after her second child was born and 12 years later was unusual in that she chose to work 30 hours rather than the usual 37 and sometimes worked at home on Fridays. It seems rare for a senior scientist/manager to work part-time in a public sector research laboratory. However, despite supposedly working part-time, she said that the human resources department had commented that she usually worked longer hours than some staff who worked full-time. Gina said she chose to work part-time as it gave her the freedom to be flexible to take time out from work to go to her children's sporting events, for instance. Arguably, Gina was being exploited by her organization but that was not how she saw it:

I work shorter hours because it gives me more flexibility. I don't want to feel guilty taking the time back sort of thing, so I thought I'll keep it as it is and then you know go to sports day and I just write it in my diary when I'm doing it and I'm not taking advantage of the system. (Gina)

Jane, a medical doctor, professor and CEO of a public sector organization, was more fortunate than most of the women healthcare scientists. Like Nicola, Mary and Rosa, Jane worked part-time for six years after the birth of her second child and then her husband made significant changes to his business career to accommodate the children so that he could work from home to complement family life. In doing so, he became the main carer of their two children:

> He said, 'Look we can't do both. You've now got a full-time job and you've got a secure salary, why don't I give up and look after the children?' (Jane)

Jane and her husband were privileged to have sufficient income to afford to pay for help in the home over many years, to nannies, housekeepers and gardeners, so her husband was able to choose the bits of attending to the house and family that he wanted to (see Hochschild, 2003; Maushart, 2003), options that were not open to most of the other participants:

> I had my children fairly late as well. I remember one piece of advice to me was either have your children very early or wait till you're a consultant. So when I had my children I was already a consultant so in a way I was secure in my job before I had the children. (Jane)

Working part-time after maternity leave did not seem to affect the course of Jane's medical career as she became a CEO of a public sector institute and had held other very senior national and international appointments.

In conclusion, because children don't fit into the world of science, women with children are silenced, as Smith describes (1987, p. 9). Women are nervous of asking 'favours' when home life encroaches into work areas, for instance if a child is sick. These are masculine norms that women avoid challenging because doing so highlights their vulnerability especially if they have children and it also indicates the fragility of their careers.

TAKING LONGER CAREER BREAKS

Most women wanted a lull in their career of about five years without permanent damage to their careers. They did not want to leave science altogether if they had children but were concerned about how they were going to balance family life with a return to a scientific career. Alice, Elena,

Angela and Mary took career breaks longer than the statutory entitlement, Angela and Mary each taking around six years.

Alice had two children in her mid-thirties when she had already begun to build a scientific reputation. After the birth of her first child, she had a number of publications but wasn't happy in her job so she took redundancy and had her second child. During this extended leave, she considered her next career move:

> I had more than 20 publications by then. I had a career break for 18 months and stayed home without a job to have a family and had my second child. I had some redundancy money and I thought about what to do next, then I started looking for a job to get back into science it took about six months to find one. It wasn't easy and the first one wasn't my ideal job but I took it. (Alice)

Alice accepted a job she did not think was 'ideal' but this turned out to be a good strategy for her as she was then able to move to her current senior position in this organization where she became highly regarded nationally and internationally. As with Gina, there were limits as to how highly she was regarded within the organisation itself by her male bosses. Most notably, she was not invited to sit on the senior management team of her organization, where there were no women members. More of Alice's story is told in Chapter 3.

Another woman, Elena, in her early thirties with a baby of 20 months, took a year at a time when nine months was the maximum maternity entitlement, negotiating successfully with sympathetic bosses to have longer leave and to return part-time to the same organization:

> I carried on until the end of eighth month of the pregnancy and then I took a year out. I decided to take a full year maternity leave. It was difficult to come back but my colleagues were great. (Elena)

However, as noted earlier, she devoted part of her maternity leave to her research, completing writing a paper with colleagues and keeping up with publications in scholarly journals so as not to feel left behind when she returned to work.

Angela was in her twenties when she took an extended maternity break of six years, so was younger than the other participants when she had her children. During this break she took on several voluntary roles at the local school, which she said 'exercised her mind':

> I did some voluntary work at the school and having children gave me a new confidence and I became more outspoken and more sure of myself. (Angela)

Angela was a good example of a woman who sought out opportunities to progress her career by gaining a PhD by part-time study when she was in her mid-thirties, after she had children. It was several years later following this absence from science that Angela began to realize she needed a PhD to advance her career. She then had difficulty convincing her medical boss of this because he considered it was unusual for a married woman with children to be committed enough to do a PhD. However, she successfully negotiated with him and after over a year of pressing him, she persuaded him to let her study part-time for a PhD, which took her nine years to complete:

> He took some persuading but he eventually gave in. I used to go and see him every few weeks with a new idea. He must have got fed up with me. I think it may have been because by then I had three kids and he was not sure of my commitment so I had to work on him quite a bit but he eventually gave in and then was very supportive as a supervisor. (Angela)

So, with perseverance Angela found a male boss who agreed to supervise her to help her achieve her aims of getting a PhD, the qualification needed for a more scientific role and a more senior scientific post:

> I was lucky I think. 'Lewis' [Angela's boss] had said that he thought women only worked for pin money but he was the type of person to make remarks to initiate a discussion; he could have been expecting a vehement response but of course because of his position people didn't often confront him, including me. (Angela)

Angela wanted new challenges and perhaps using this knowledge, her boss chose to make her redundant rather than one of her two male colleagues who were on the same grade. She then applied and accepted another post moving over a hundred miles. In the new organization, within two years following the retirement of the head of department, she was promoted to his post and sat on the organisation's senior management team. Angela was the only one of the four women who took an extended career break who was able to reach that level and one of five women overall who acted on a senior management team, indicating an important achievement. She was also able to develop her scientific career becoming recognized nationally and internationally.

Mary, like Angela, was active in a variety of the non-scientific voluntary roles during her six years out of science to concentrate on looking after her children. During this break, her work on the voluntary projects extended her knowledge and skills base. Like Elena, she said that she put her family first at the time when she had her children. Mary had two children in her thirties when she left her previous organization and did some unofficial work for them during her first maternity leave but did not expect to return to her original paid job. Unlike Elena and Angela, Mary said she never expected to return to science but to make a change of career direction into teaching, rather like Nina was now considering. However, she sought and gained a part-time temporary job in her current organization when in her early forties after the six-year break, and resumed being an enthusiastic and energetic scientist. However, she had significant difficulties in being accepted as a scientist because of her sex, age, having children, working part-time and being a registered biomedical scientist rather than a recognized research or clinical scientist (more of Mary's story is discussed earlier in this chapter as she returned to work part-time).

MANAGING A CAREER WITH OLDER CHILDREN

As children get older, managing a home and a career took a different kind of planning and organizing. Rosa still needed to juggle work and home life and it wasn't any easier despite her children now being older teenagers, just different:

> You wish that you could perhaps stay a bit longer just to finish what you're doing rather than having to be disjointed and do it elsewhere at home and then you've obviously got the family pulling you as well, feeling that you want to do a good job bringing them up. (Rosa)

Rosa took work home with her that she couldn't get done during the working day after taking her children to various classes and cooking the evening meal:

> Nearly every night of the week one or other of them [her children] has a music class or rugby or whatever and I have to take them and pick them up. Then I go home and cook the meal, and then I'll try and finish off what I haven't got done in the day. We're getting ready for another accreditation visit at the moment and I can't concentrate at work. There's too much going on and everyone wants me to do what's important to them. (Rosa)

Rosa was at that period in her life when her children were about to leave home and when she reflected on her career achievements she found them wanting and felt frustrated:

> I'm frustrated because I feel as though I'm neither one thing nor another at the moment. (Rosa)

Rosa's frustration was seemingly due to the conflicting pressures on her from her family and her career and it seemed she was worried she was failing on all counts. She was working the second 'shift' described by Hochschild (2003), looking after children and managing a career.

Carla, on the other hand, found what seemed an unusual but apparently effective way of coping with having a child and a career:

> Actually what I do is compartmentalize. When I walk out of that door, I don't think about work. I think about picking 'Josh' (Carla's son) up from school and what I've got to do, homework and everything, I don't even think about work at all. When I'm at work, I don't think about Josh or my family or anybody else in my family, not just Josh, when I'm at work, I don't think about them. That area has to be switched off; that's left then and I've always done that. (Carla)

Carla had recently had an operation and was obliged to take several weeks off work, which had given her the chance to be a 'proper mum':

> When I had my op I was off so long that Josh got really good at his tables and spelling so I felt really guilty about going back to work because he'd progressed so much and enjoyed me being at home so I did feel guilty when I went back. I liked playing at being a proper mum. (Carla)

By saying she was 'playing at' being a 'mum', Carla indicated that most of the time her parenting role took second place to her career on which she was very focused:

> I switch off when I leave him behind. I've always done that, I mean apart from you know the first few days when he was in the crèche when I was heartbroken but you know normally I'm not the sort of person that's going to have all my family pictures round my computer. It's like I'm at work now and I'm this person. (Carla)

Carla here implies criticism of people who 'bring' their families into the workplace by placing photos around the work environment. Despite saying she stopped thinking about work when she went home, Carla sometimes worked after her son had gone to bed:

> It may be that after Josh is in bed it may be then that I may have to do more work in the evenings but I try not to as at home I'm a mum then and it just has to be compartmentalized. (Carla)

Like other women, including Rosa discussed above, Carla undertook this additional work but it was hidden work and not noticed. However, she emphasized that she was able to compartmentalize her life into more than two parts, differentiating when she went out to a social event:

I don't have very much social life because you can't fit that in really other than occasionally at the weekend. If I'm going out and I've left Josh with the babysitter then I'm then another person again, I'm a person having a drink and having fun I'm not a mum and I'm not a scientist either, I'm a different person and I think you almost have to dress the part and behave the part and have to be all these different people. (Carla)

Carla was notable in saying that she manifested different identities depending on the situation she was in at any one time. It is interesting to consider that she was portraying multiple identities when being interviewed for our study.

PUBLICATION RATES AND MOTHERHOOD

A reduction in the numbers of publications during motherhood disadvantages women scientists (see Posen et al., 2005). Several pieces of evidence submitted to the HoCSTC report (2014) commented on the way women are disadvantaged and the report, citing evidence submitted from the Science and Technology Facilities Council, Women in Science, Technology, Engineering and Maths Network (STFC WiSTEM) states:

[A]ssessing publication records by the number and impact of papers produced 'militates against career breaks or reduced working hours'. For example, the h-index,[1] a commonly used measure, makes no allowance for time away from research or for part-time working. (HoCSTC, 2014, p. 33, citing evidence from the STFC WiSTEM Network)

Nina, in her early thirties with a PhD and about to have her second baby, was unable to see how she would be able to make the grade as a senior scientist when she would not be able to publish significantly during the five years when she wanted to concentrate on her children. Nina was very aware that her career relied on publications as a key marker of success (Delamont, 2003; Greenfield, 2002a; HoCSTC, 2014; Hosek et al., 2005):

I've not been able to find any advice for female scientists who are taking a career break or that sort of thing. Your publication record is so important. I'm not even sure how taking a break to have children will affect my career in the long term but it can't be good can it? (Nina)

Harriet raised similar concerns as she knew her publication rate had reduced since the birth of her son and was well aware of how she would be judged:

They [publications] sort of peaked just before I had my son and then slowed down. It really does slow right down. As that is one of the main criteria by which your success is measured, you have to work hard to build them up again. (Harriet)

The HoCSTC report (2014), accepting evidence from Cardiff University, wrote that there was 'pressure on women to "come back as early as possible" [to] keep up their publication output or "research productivity"' (p. 37) and Lisa, who had taken very little maternity leave commented:

Not only would I let down my students if I'm off but if I don't publish, my status as a scientist will go down and I don't want that. (Lisa)

The importance of publication rates not only influenced promotion prospects but also influenced how scientists are regarded by their peers. Ralph, the chief executive of a public sector organization, commented on the publication rate of a woman lecturer when he was a student, linking it with her status and remarking that he had no female role models:

My supervisor in biochemistry was a woman but she wasn't one of the high flyers in the department in terms of publications and status. (Ralph)

He stressed the importance of having a good publication rate himself when he was considering changing career direction:

I felt a reasonable degree of confidence at that point because my research work was going pretty well and I had a decent lot of publications. (Ralph)

The notion that reduced publication rates due to maternity leave could be taken into account at the recruiting stage for scientists had not been considered by Ralph: he had never thought of making any allowances in differences in publication rates when interviewing women scientists who had taken maternity leave, remarking:

Well, I confess I've never really thought about that. I've always regarded a publication as a publication without thinking about the personalities behind it. (Ralph)

Significantly, Ralph chose the word 'personalities', which seemed to be an unusual use of the word in this context, rather than a word such as 'reason'. In doing so, he apparently preferred the association of pregnancy with personality and thereby avoided linking publication rates with the possibility of a woman scientist being adversely affected because of maternity leave.

Rather than discuss maternity leave further he moved on to talk of the 'prolific publication rates' of the women scientists in his institute, possibly to avoid further confronting an unpalatable notion of his responsibilities to women in his institute who might be disadvantaged in terms of their publication rate when they had children.

Although, at the time, there was a hint that Ralph might be interested in considering issues related to assessing women's achievements differently in the future, two years later Ralph still had made no female appointments to his senior management team, which remained all male.

The pressure on women to maintain their publication rates during and post-maternity leave could be reduced if publications were assessed differently, perhaps over a shorter period of the applicants choosing or limited to a specific number of key publications. Alternative ways of assessing the contribution of applicants could also be considered, as the STFC WiSTEM Network proposed in their evidence to the HoCSTC report: 'quantitative measures of staff and job applicants' productivity such as number of papers published and *h*-index should be replaced with a comprehensive evaluation of the person's contribution to the organization and the field' (HoCSTC, 2014, p. 34, quoting the STFC WiSTEM Network). They suggest that credit should be made for more relational activities such as mentoring and counselling.

LONE MOTHERS

First, some statistics: lone mothers' employment rate increased from 43 per cent to 60 per cent between 1996 and 2013, corresponding with 'a fall in the percentage of women remaining out of the labour force to look after the family and an increase in the percentage of mothers in employment' (ONS, 2013, pp. 4–5). Ninety per cent of lone parents are women (ONS, 2013, pp. 9–10, figures for April–June 2013). The percentage of lone mothers in employment in the UK is lower than for mothers with partners: 39 per cent of lone mothers with a child under three are in employment compared with 65 per cent of women who are in a couple; 61 per cent of lone mothers with a child aged between four and ten are in employment compared with 74 per cent of women who are in a couple; as the child(ren) get older (11–18), the numbers of lone mothers in paid work and mothers with husbands or partners narrows (74 per cent for lone mothers and 80 per cent for mothers in a couple (ibid.). The gaps overall have narrowed between 2008 and 2013 but the figures reflect that it is more difficult for women with younger children to obtain paid work, whether due to the needs of mother and baby or to the availability of work.

Statistics from the ONS indicate that most men with children work and the men who bring up their children alone are few compared with the numbers of women on their own with children (ONS, 2016). Even though pressure comes from successive governments for women with children to work, looking after other people's children is poorly paid and women who undertake such childcare are often exploited (Ehrenreich and Hochschild, 2003). At least one public sector nursery closed (at the Health Protection Agency) due to its cost to the organization, obliging women and men to make other less convenient arrangements for childcare.

Nevertheless, in contrast to the complexity of the responsibilities and the lives led by the women who had both children and who had husbands, it could be regarded as surprising to find that the three women (Carla, Harriet and Deidre) with dependent children who did not cohabit with a husband or partner at the time of the interview, described notably well-ordered lives. Chandler (1991, p, 123) sheds some light on the well-organized lives of these lone parents: 'Women without husbands have a great sense of their own self-sufficiency and independence'.

Furthermore, Charles and Kerr (1988) point out that women often look after men by paying attention to their needs in ways that women don't bother to do for themselves. For instance, women often produce 'proper' meals for a 'proper' family because of their husbands, and women often omit to prepare such meals when they only have children and themselves to feed (Charles and Kerr, 1988, p. 227; Weiss, 1979). This phenomenon is not new. In Hetty Morrison's book in 1878 [2010], she comments that she had heard her neighbour's following observation many times over the course of her life (p. 35): 'I am going to have an easy day; my husband will not be at home, and I never do any cooking just for myself and the children'.

Whilst men may also omit to cook for themselves when on their own, it is generally women who take responsibility for meals in the family home. Arguably, husbands create more work for their wives, as Maushart (2003) remarks about the time when she was a lone parent with three children: 'I had a great less work to do without a husband than with one . . . as . . . McMahon has observed, "the help that husbands provide does not even cover the amount of work they create" (p. 28, quoting McMahon, 1999).

Charles and Kerr (1988) point out that with less money coming in and with no one to share the burden of the home work, life may be more difficult than where two parents are present, but this did not seem to be borne out with the three women interviewed. In part this could have been because their children were beginning to grow up and the difficulties the women may have faced when the children were young were now reduced. Nonetheless, their lives seemed remarkably well organized.

Carla, in her early forties, had a partner but did not live with him and appeared very well controlled both in her work and home life. Throughout the interview, she talked quickly and her confidence about herself and her work role was clear: 'I want to move things on in a positive way'. She had left two posts in the past because she could not see any opportunities there and did not want to work anywhere where she would feel bored. Carla had 'fought' for her position in her organization for which she expected to be 'rewarded' both financially and by an increase in position. She raised her own profile by 'put(ting) my head up above the parapet'. Carla completed her MSc before having her son (now eight) and had taken four months maternity leave. We describe elsewhere how her boss prevented from continuing her PhD after one year's study and at the interview some 15 years later, she said she was considering studying for a Master in Business Administration (MBA). When this was followed up a further five years later, Carla had not started studying for an MBA. It seemed that she had settled for working in a job she found satisfying and further study was not on the horizon.

The second single mother, Harriet, was 40. Harriet had finished her PhD ten years previously during her pregnancy and took seven months maternity leave:

> I've enjoyed my career. I feel I am very lucky to come to work and actually want to come to work and there are things I want to get done during the day. I think there's a very small proportion of people who can actually say that they're happy to go to work. (Harriet)

Harriet had support from a 'fantastic childminder' and in cases of difficulty said, '[M]y dad finishes work early and he'll pick my son up'. For a short period things were less easy:

> At one point I had to drop him off at 8 o'clock in the morning and pick him up at 6 o'clock in the evening and you know I hardly saw him when he was awake. (Harriet)

Harriet was fortunate that her son was rarely ill. She was also fortunate that on the few occasions where he had been sick, her sympathetic (female) boss, encouraged her to work from home, writing reports and papers and so on, rather than taking annual or carer's leave. Harriet took both paid and unpaid leave to manage school holidays and her son would also spend time with her father and with the 'fantastic childminder':

> I'm still in touch with her. He would stay with her till 2 o'clock and my dad works funny hours so he will finish work and pick my son up. (Harriet)

Her approach to her career was the wish to be able to look back when she would retire with satisfaction:

> I think it was a case of thinking what if I look back when I retire and I don't like what I see. I want to be happy or content with my career and to spend all the time that I can with my son as well. (Harriet)

Harriet accepted that her career could have progressed faster as she had been obliged to reject opportunities because she had her son to think of:

> I didn't have the energy to rise to a challenge. I could do my job but I couldn't go above and beyond that point which I think was probably required to get on. (Harriet)

Having a child delayed Harriet's progression but it didn't stifle it. She talked about times when life was difficult and used this to explain why she had not progressed further in her career:

> I have to say I think I've had opportunities especially when my son was younger when I could have pushed things forward. I mean, you'd expect in your mid-thirties, you know when your brain's operating most efficiently, that perhaps there were opportunities then and I did turn things down, but if you're going to grab hold of an opportunity you've actually got to put energy in but I just really didn't have the energy or motivation to take them up at the time. (Harriet)

Harriet saw it as her own responsibility to 'push things forward' and didn't ask for what she saw as favours:

> I would take annual leave if he was ill, although sometimes I'd take work home like reading papers or writing reports. I was lucky that I could do that. 'Josie' (her boss) didn't say, 'No you can't', but I do know of people who couldn't. I would be very careful about not asking too often or asking for favours. I wouldn't want to be labelled as someone who couldn't cope. (Harriet)

Harriet was lucky compared to some female colleagues who were not able to take work home when their children were ill. But she was isolated as 'm[o]ther' and was concerned about asking too much of her boss for fear of being labelled as a woman and mother who could not deliver on her scientific demands so she kept a low profile.

Unlike Carla, Harriet was less overtly confident and said she once became upset when somebody suggested she was ambitious, initially thinking it was acceptable for a man but not for her as a woman. She reviewed these thoughts, however:

> Ambition is seen as a bad thing in some ways you know. I've been accused of
> being very ambitious and I took offence at it and then I thought well why should
> I worry about that; ambition can be a positive thing. (Harriet)

Harriet was concerned at first that being ambitious would contradict being
feminine and being a 'nice woman'. She recognized that she could begin to
move forward again in her career, now that her son was growing up:

> I am beginning to start doing more committee work and I do see myself as a
> deputy here and need to be involved in lots of different things. I have my own
> areas as well so I feel ready but I can't see myself leaving the science side, I enjoy
> it far too much so I'm now looking for bigger projects. (Harriet)

It was of little concern to Harriet that she did not live in an 'ideal type'
family (Weber, 1949):

> [. . .] where the husband would be greeted by his lovely wife with the apron on
> and the children and they'd kiss him and then they'd go to bed and then he'd
> sit down with his pipe and stroke the dog in front of the fire while his wife, you
> know, fetched and carried for him – well not in my house. (Harriet)

Although Harriet apparently considered the traditional family as 'ideal',
she also viewed it with irony, as unobtainable and dispensable.

By contrast, for Deidre, her salary as a medical doctor made life easier
than for Carla or Harriet. She was divorced and now without a partner
and with two dependent children but she coped by earning a high enough
salary to employ live-in 'help':

> I've still got full-time help because you know there are quite a lot of interna-
> tional things I go to and you need somebody there and I have help in the garden
> as well. (Deidre)

There are parallels here with the executive who has the stay-at-home wife,
as described by Hochschild (2003) and Deidre was thus paying for the
input from a substitute 'stay-at-home wife', which many men expect as an
entitlement and for which they also 'pay', but only in supporting a wife,
not as an additional salary. Medically qualified doctors are in vocational
careers and Deirdre chose an area of medicine in a research institute where
she did not have to work out of hours and that fitted in well with having
a family.

Interestingly, much of the literature on lone parents (Chandler, 1991;
Charles and Kerr, 1988; Weiss, 1979) does not seem to tackle the issue of
lone mothers as career women. The emphasis is on relationships, and in the
Weiss's (1979) case, how to cope with being 'different', a word Weiss uses

frequently and also pervades this research in relation to women generally. In addition, authors who discuss the issues of women with children and paid work seem to concentrate on two-parent families (Delamont, 2001a; Gatrell, 2005, 2008; Hakim, 2000; Hochschild, 2003; Maushart, 2003; Martin, 1984; Potuchek, 1997; Williams, 2000). Yet the emphasis from the UK government has been on the drain on national resources by lone parents, and persuading the single parent (by implication, working class and young) to do (paid) 'work' but no mention of developing a career.

For instance, the Freud Report (Freud, 2007), an independent report for the UK Department for Work and Pensions on the Welfare-to-Work strategy, suggests options for reducing the dependency of individuals on welfare benefits and encourages women to look for work. The lone parent employment rate is quoted as 56.5 per cent (p. 9) and recommends that the 'Government reduces the point beyond which a lone parent can claim income support from when their youngest child is 16 to 12' to help achieve the government aspiration for an 80 per cent overall employment rate (p. 91). The government accepted the recommendation, and from 2010 lone parents have been required to look for work when their youngest child reaches 12 years old.

The target for this change in legislation is apparently women with low incomes, not lone women with careers. There is no emphasis on getting women into careers, merely 'work'. One interpretation is that it is believed to be difficult enough for married women in two-person households to cope with a family and a career; single women with careers and children are so few that they may be ignored and discounted.

According to Chandler (1991), 'In conventional marriage, domestic routines are fitted around male schedules and career patterns' (p. 118). The three women in the study who were lone mothers probably benefited from having 'uncontested' control of the household and not having to consider a man's needs (ibid.). Unlike Deidre, who seemed little affected by her status as a single parent and who probably benefited from choosing a career in medicine, the careers of both Carla and Harriet would have been likely to have progressed faster, had they not had children. This seems to be similar to most of the women with children who were interviewed and was probably not related to their marital status.

MEN'S VIEWS OF MOTHERHOOD AND FATHERHOOD

With the exception of Alec (whose story is told in the next section), family life did not figure much in the interviews with the men. Where it was commented upon, it was generally quickly passed over. It seemed that the men did not worry too much about their families, probably because their wives did all the worrying for them.

Larry commented: 'I think family isn't a concept I've thought about much – I suppose it only happens when kids come'. Bob said he worked long hours during the week so that he could spend time with his family at weekends and that the decision for his wife to give up work when they had children was one 'jointly made'. And Frank, from the point of view of a chief executive, said 'we try and encourage various aspects of flexible working, I think you know, job shares and part-time'. Tim said that the decision for his wife to give up her science career was one made by her:

> 'Sue' made the decision to stop working for five years; she wanted to do that and give up the science because she realized that if you stopped for five years it was very difficult to get back in and she wanted to do teaching. (Tim)

Few of the men interviewed commented on the prospect of having children. Ben, a 34-year-old male scientist, when pressed, considered the implications for him of having children:

> I wouldn't mind being at home with a baby or a child, I wouldn't say forever but I'm quite good at filling my time. I think my balance is not making a choice or conscious choice, not deliberately. I wouldn't say that I don't want children but I don't particularly want them right now. I'm not thinking I must go for my boss's job even though I can see it's a logical progression and I think by not consciously having made choices at the moment anyway it's perhaps easier, but then it may be harder when a choice has to be made. I think it would depend a lot on whether my wife would be working, I think if she is where she is now maternity would be quite easy, but if she moved jobs, it could be more difficult for her to take a break. (Ben)

Interestingly Ben didn't consider that looking after a baby would be a full-time endeavour and thought he would be able to have time to do other things other than look after a baby. He was in the position of not having to make a decision yet, so, at 34, his hypothetical thoughts were not under pressure of a maternal clock, at least for himself. He was the only one man of the male interviewees who was thoughtful about maternity leave or the pressures of having children:

> I have never thought about the point of view of a women, I have never because until you experience it yourself you can't understand what it might be like; everybody's life is different. I'd say the choices that women make, I'd say the parenting aspect probably preys on their minds more, they are thinking about their next move and how to structure a career, I would imagine it plays more of

a bigger part of their thinking then it does a man but I don't know, that would be my guess. (Ben)

He remarked that the women in his organization seemed to take most of the responsibilities for children:

A father of two young children here would only leave early if his wife couldn't make it. Other than that there are a couple of women who have got young children and they tend to be the ones who leave at three to collect them from school, so they work to flexi time. (Ben)

These few comments affirm that it is expected that women take on most of the responsibilities for the children. Most male interviewees seemed content that this remained so and for women to retain responsibility for domestic care agendas. Alec's experiences were very different, as we now go on to report.

EXPERIENCES OF A MALE CARER

One man, Alec, from the cohort of the men in the study, was unusual as he had taken on the role of joint primary carer with his wife for their two daughters (see Gatrell, 2007 for a discussion on co-parenting; Coltrane, 1996). Alec's decision was in keeping with an increasing minority of men who seek to be closely involved with children's upbringing, especially when they are infants, either through desire or necessity – for example, because partners are also in paid work. In the past, men in intact adult relationships might have been satisfied with maternal negotiation of father–child relationships (Ribbens, 1994). However, in the years since 1990, attitudes among fathers have shown a change. Some men actively seek to reinterpret their paternal role from that of main breadwinner to engaged and involved parent (Beck and Beck-Gernsheim, 1995) and some men demonstrate increased child orientation in the context of both marriage/cohabitation and divorce.

There were a number of consequences of Alec's decision to take on this prime role with his children, including reducing his hours to cope with the additional family responsibilities to enable him to pick the children up from school and care for them while his wife worked longer hours. Gatrell (2005) describes 'Charles', a scientist like Alec, who found that it was not acceptable in his science organization to ask for flexibility: 'And it's bad enough for mothers who ask for flexibility in their work, they will sense the whiff of disapproval, but for a *man* to do this – well that is completely beyond the pale' (p. 203; emphasis in original).

Alec, who was a senior scientist with a PhD, didn't know what he wanted to do after his first degree and started work as a school laboratory technician because the job became available. He changed jobs several times, married, and his wife noticed an advertisement for a university research assistant. The job involved setting up a new laboratory and came with the opportunity to do a PhD, which Alec grasped. He spent the next seven years working and studying, sometimes in his own time and during that period they had their first child. He took a promotion to work in his current post in another organization where he had been for ten years.

Alec described several instances where he suffered prejudice in his current role because he did not fit the normalized male senior scientist role and considered that this had had an adverse affect on his career:

> I mean I've got a PhD and they call me a senior scientist here but I'm not because I'm still two steps down from where I should be and the reason is because this childcare business has thrown it. (Alec)

Alec considered that, as a man, he had suffered more discrimination than a woman would in his public sector organization. He felt that it was not accepted that a man might want to work part-time or to need to leave work in an emergency to accommodate children's unplanned and urgent needs:

> I got comments like what's your wife doing [and I'd say] well you know she's my child as well. (Alec)

According to recent research by Gatrell et al. (2014), Alec was not alone, either in his experiences of balancing childcare with work, or in his views about the relative ease for mothers in gaining access to flexible working. As regards difficulties gaining flexible hours, Alec's experiences were in line with current research showing that line managers often resist offering work–family benefits to male colleagues, due to assumptions that fathers neither want nor need these (Burnett et al., 2013; Gregory and Milner 2009, 2011; Holter, 2007; Tracy and Rivera, 2010). 'Deeply ingrained' assumptions about masculinity at work may make it difficult for fathers to balance paid work and child care in practice (Holter, 2007). Lupton and Barclay (1997), Miller (2005), Dermott (2014), Featherstone (2009) and Gregory and Milner (2009, 2011) all identify the challenges faced by fathers trying to overcome employers' assumptions regarding what is appropriate paternal behaviour for fathers.

Consequently, in keeping with Alec's situation, the fathers' experiences in contemporary workplaces suggest that policies promising flexible working may in practice be difficult for men to access, with senior colleagues finding reasons for blocking men's right of entry to family-friendly initiatives. Alec related instances where more senior women than he were able to reduce their hours of work, including working term-time-only contracts and had not suffered for it:

> They have a number of women in this place who are far more senior than me but they are all on reduced hours and term time working to fit the childcare round it, but they didn't like me doing it. (Alec)

Alec's assumption that women colleagues find it easier to obtain flexible working than men in equivalent positions also accords with the views of other fathers in similar circumstances. Fathers in Gatrell et al.'s (2014) research indicated their belief that while flexible working policies were apparently offered to 'parents', these were often, in practice, aimed only at mothers. Like Alec, fathers in Gatrell et al.'s research felt that their access to flexible hours was compromised by maternal privilege. They presumed that mothers had the priority as carers and that mothers must therefore be able to easily gain permission to work flexibly. In practice, it is worth noting that these fathers' opinions about mothers and flexibility tended to be envisioned through rose-coloured spectacles. Mothers in Gatrell et al.'s (2014) research who sought flexible hours experienced, variously, refusals, downgrading and marginalization. Perhaps the key issue here is that any worker who seeks to work flexibly in order to care for dependent children will be regarded by employers as disruptive and less committed than full-time employees.

In addition to his concerns about constraints on his opportunities for accessing work–family benefits, Alec also considered that working reduced hours had unduly influenced his possibilities for promotion. His career was constrained like those of the women and he found that the discrimination had continued despite returning to full-time employment:

> I think there's a problem with men putting their family before their work you know before their career. They don't like it here and my career has suffered and it's certainly affected my career prospects. (Alec)

Alec considered he was treated adversely compared with the women in his organization who had children (Gatrell et al., 2014) and in the above quote distanced himself from the organization he resented by using the third person 'they'. But, by working part-time, Alec considered his relationship with his children was much closer than when he worked full-time:

> Some men have a career and they become almost tied in and they can't give it up but then my feeling is a lot of them are not as close to their children as I am

close to my girls. I have a much closer relationship with my daughters than a lot of people I talk to have. (Alec)

Alec had taken on a role usually filled by mothers and he found similar difficulties to the women. The way he juggled his children and work was much the same way as the women in the study said they managed their complex lives. As Alec saw it, however, he suffered more discrimination because he was a man playing a role expected of a woman and he considered that his employers did not accept his right to put his family first on occasions (see Gatrell, 2005).

Beck and Beck-Gernsheim (1995) suggest that children have become more important to fathers because they have more 'permanence' than marriage (p. 73) but this did not seem to apply to Alec's shared role where he appeared very committed to his wife and daughters. In his description of his role at home, Alec concentrated on the benefits to him and his children and he certainly had invested considerably in his children, to the detriment of his career. It did not seem that he viewed parenting of his daughters differently from that of a mother or that the way he balanced his children and work was so very different from the women. He was, however, very bitter about his lack of career prospects, and about the attitude of his organization, which he felt restricted him and reduced those prospects. It was not clear, however, that Alec actually suffered more than a woman would have done in similar circumstances. But, as a man, it was not an experience that he had expected to endure and he did not find the situation acceptable and felt bitter and resentful. By taking on a role usually associated with women, Alec was treated more like a woman at work, alien and different, and he didn't care for it (see also Gatrell et al., 2014).

As early as 1993, Kimmel (1993) suggests that organizations need to be more overtly inclusive of fatherhood within their work–life balance policies. Kimmel argues that both fathers and mothers should be facilitated in balancing parenting and paid work without this negatively affecting their careers. If science-based organizations could be more flexible in their treatment of involved parents, women might be less likely to be branded unfairly as disruptive and less uncommitted once they become mothers (Kimmel, 1993).

SOME CONCLUSIONS ON PREGNANCY, CAREER BREAKS AND CHILDREN

Balancing pregnancy and working in science was not easy for the women in our study and it is likely that the return to work part-time and long career breaks contributed significantly to their lack of career progression and failure to reach their maximum potential. Despite discrimination in the workplace against women (whether pregnant or not) being illegal (Equality Act, 2010), bosses found subtle ways of keeping the women in their place. Comments suggesting that the women didn't need to rush their careers but then saying that further training was not appropriate for them in the meantime, and a lack of support for maternity leave and to cover work during the women's maternity leave were common. To progress their careers after having children the women needed perseverance and negotiating skills to overcome the barriers put in their way. Sometimes the women overcame them and sometimes they could not.

Whilst negotiating skills were apparent in some women such as Angela, they proved to be difficult for many women and they kept a low profile, as Miller (1986) notes. Several women found it very difficult to put their careers in science to one side when they had children, however highly they valued motherhood. The situation didn't get much easier when the children were older. Most women coped and continued with their multiple responsibilities but were pulled in many directions. Quite apart from the emotional pressures of leaving a baby in the care of someone else, there were very practical issues of organizing childcare together with the issues of having a responsible job, and not wanting to highlight their own dilemmas at work, preferring to remain invisible and keep a low profile. Some women kept quiet about their pregnancies for the same reason.

There were some common factors in the obstructions the women met. Some of the barriers related to the need to catch up with others who had not made a career break, but the reasons for the delays and setbacks go much deeper. The women did not want to highlight their own situation because it would draw attention to the very obvious sign of being a woman and sexual being (Martin, 1990), apparently preferring to disregard or even hide their pregnancies initially, as Nina had done. Part-time working and taking extended maternity leave seemed to signal to some bosses that they lacked commitment to science; pregnant women, and women who take extended periods of maternity leave, and women who return part-time seemed to become labelled as 'm[o]ther' rather than 'scientist'.

There was a wide range of achievements from the women returning to working in science part-time. At one end of the spectrum, Jane, a medically qualified doctor, had reached the position of CEO; Gina rose to being a national and international expert (although did not sit on her senior management team). Nicola, Nina, Rosa and Mary were at the other end of the spectrum. Their wish to keep a low profile and not to negotiate their career development with their bosses and organizations helped keep them in their place – the place of the operational woman scientist. Working part-time was just one of a number of gendered barriers that highlighted their differ-ence and affected the career progress of the women detrimentally.

Angela and Mary both took extended career breaks of about six years following having their children. Angela studied for a PhD afterwards, worked hard to become a director of a laboratory as well as retaining her scientific expertise as a healthcare scientist, was recognized nationally and internationally, directed a laboratory and sat on her senior management team. Mary, on the other hand, was unable to persuade her boss to let her study for a PhD and she eventually left science altogether. Some bosses provided patronage or advocacy and several women were helped to pro-gress by such assistance, but such support was insecure and could be subtly removed, which was severely detrimental for those women.

As Babcock and Laschever (2003) describe, the process of negotiation is gendered in favour of men and, for most women, the lack of negotiating skills and not wanting to disrupt the status quo hindered them on account of their gender. Assessing their career trajectories the women exhibited limited autonomy and were constrained by the norms and sanctions of the science organizations in which they worked, by professional demarcation barriers, by bosses and by society, which positioned them in their subordi-nate 'place'.

In keeping with Gatrell's (2005) and Marshall's (2000) research, several women, including Mary, Nina and Elena, were considering moving to new jobs, some away from science altogether, rather than continuing the exhausting challenge to their existing bosses to help them pursue careers in science because they were labelled as m[o]thers. Mary, for instance, took up scientific writing to move away from the subtle masculinities at work that blocked her paths to advancement as a research scientist. Some women left science completely because they found juggling a family with a scientific career too much. At the time of the first interview with Nina, she was considering leaving science to train as a teacher, as two of her colleagues had done recently and four years later she did just that so being a loss to science. Elena's possible solution to her dilemma of combining science and a child was to change careers to be a translator, a skill she learnt during her maternity leave but by so doing she would also be lost to science.

These non-medical women were isolated by their 'difference', which kept them from the place of excellence and science. They accepted their own place in the masculine world of science that they were used to as the norm. Their opportunities to negotiate for advancement and their chances of actively exercising 'choice' were low. In addition, the women were labelled as 'm[o]thers', rather than scientists and endured the double deviance described by Laws (1975, pp. 53–4) of being often labelled as a support worker and 'm[o]ther' rather than scientist (see also Cooper, 1992; Fotaki, 2011).

A drop in publication rates was a concern to all women scientists and we welcome the recommendation from STFC WiSTEM Network in their evidence to the HoCSTC report (2014) who offer a part solution that would be by replacing the quantitative assessment of applicants' publications with a more inclusive and wide-ranging assessment of their wider contribution: 'For example, they should provide acknowledgement and credit for tasks such as organization of group seminars, engagement with visiting school-children, mentoring junior colleagues, taking on placement students, acting as counsellors' (HoCSTC, 2014 p. 34, quoting the STFC WiSTEM Network).

We also support the European Council Directive (2010) (see also Business Innovation and Skills, 2012) and UK government policy (UK Government, 2015) for 'shared parental leave', which came into force on 5 April 2015 to allow more flexibility for couples whereby eligible mothers, fathers, partners and adopters may choose to share time from work after their child is born or adopted. Parents can now share leave, opt to be off work together and/or to take turns to look after their child (ACAS, 2015; UK Government, 2015). While the notion of parental leave is a major improvement on the present situation, it is important to acknowledge that this does not apply to all groups (for example, self-employed workers would usually be exempt) and that it is UK legislation (with similar arrangements throughout Europe); the United States, for example, has no such provision. Unfortunately, as the HoCSTC report (2014) predicts, 'simply introducing a new system will not in itself change workplace attitudes towards maternity, or the difficulties caused by taking parental leave' (pp. 51–2), and one year later in April 2016, the take-up by fathers is reported to be very low at less than 1 per cent (My Family Care and Women's Business Council, 2016). The effects of the planned 'Brexit', of the UK leaving the EU, especially with regard to diversity issues, are as yet unknown.

Increased flexibility and interchangeable maternity/paternity leave entitlement should help couples such as Elena and her husband to have more options and could significantly help women in pursuing any career, not just in science. Currently there is no guarantee that men will choose to share the entitlement (though there are indications that men find it difficult to access flexible working; Gatrell et al., 2014). Shared parental leave can only reduce the discrimination against pregnant women if men as well as women get used to taking their entitlement. This would have the potential of reducing discrimination against women who might become pregnant (see Davis et al., 2005; EHRC, 2016a; Gatrell, 2006a, 2008; Pringle, 1998; Wajcman, 1998; Williams 2000).

NOTE

1. See *h*-Index, accessed 22 June 2017 at http://en.wikipedia.org/wiki/H-index.

7. Concluding remarks and recommendations

CONCLUSIONS

As we reach the end of this book, some of our conclusions are striking. First, despite many years of equal opportunities legislation and proactive initiatives such as Athena SWAN, career trajectories for women in healthcare science remain limited compared with those of men, and women continue to be allocated a subordinate place within the hierarchical healthcare science structure. We observe how (as in Tracy and Rivera's 2010 study), and despite stated intentions to effect change, there have been minimal shifts over the ten years while our research was being undertaken. Outdated approaches and attitudes of directors, senior managers, senior healthcare scientists and medical practitioners with scientific and management responsibilities to improve how gender is dealt with in science remain. In addition, assumptions that women are responsible for the domestic care arrangements in the home remain largely unaltered.

This is the case despite over 40 years of equal opportunities legislation and much rhetoric in a rich array of different government publications and other initiatives on the topic (Equalities Review, 2007; Equality Act, 2010; Greenfield, 2002a, 2002b; HoCSTC, 2014; Kirkup et al., 2010; OECD, 2012; UK Government, 2015; WWC, 2006a, 2006b). We recognize that it could be argued that this situation is not unique to women in science. As Blau et al. (2014) observe, women in a range of occupations and sectors experience limited career advancement, and earn less than equivalent men (see also Stead and Elliott, 2009). Thus, for example, Mariana Fotaki (2013) observes how women in academia may be marginalized and excluded from research opportunities and Kumra and Vinnicombe (2008) describe how women in professional services firms are not advancing to the levels anticipated when they embark on their careers and women are still in the minority on corporate boards.

Yet certain factors are perhaps particularly common to women in science. For example, in a profession that requires very specialist and complex training, it is of concern to learn that young women in science are still experiencing disadvantageous treatment and receiving inappropriate advice about becoming research scientists. Such poor advice is offered perhaps on the basis that some advisors and family members, still, anticipate that women are unlikely to become leading research scientists and do not, consequently, require certain qualifications and experience to prepare them for this.

Acknowledging the challenges faced by women in other professions, we further suggest that the hierarchical nature of the professional and other structures in healthcare science means that there are many barriers at which women's career advancement may stall. At the heart of our argument about the reasons for women's stalled career progress in science compared with men, we propose the concept of women's lower 'place' in comparison with male colleagues. We suggest that reinforcement of such a 'place' in science perpetuates the status quo.

For women with children, it seemed to make little difference at what point they had their children, whether early on or much later in their careers. The women who progressed seemed to be those who faced their bosses with persistence but without antagonizing them. The most beneficial factor for those keen to pursue a career in research science was to obtain the postgraduate qualification of a PhD before they started in paid employment. For the late developers and those who wished to be acknowledged as research healthcare scientists when they were already employed in healthcare science, it was very difficult for women to study for a PhD: the opportunities that seemed open to men were not available to most of the women. Those that progressed needed the support of a medical or senior scientist advocate and usually these benefactors were male. When this advocacy ceased (as it invariably did) the women needed to move on to find new positions to further their research aims. We have heard from professional colleagues that training funds are diminishing under the pressure of reduced budgets; thus, the opportunities to progress in research healthcare science will become even more difficult for women.

We suggest that women's place is firmly preserved at the lower end of the hierarchical professional structures in healthcare science and we have identified in earlier chapters four mechanisms that contribute to such stabilization: subtle masculinities, secret careers, the notion of creative genius and women's potential for reproduction.

In order to illustrate the effects of the four mechanisms on women's 'place' in science, we developed the framework described in Chapter 1 (reproduced here as Figure 7.1 for convenience), which helps to summarize and visualize the contributing mechanisms:

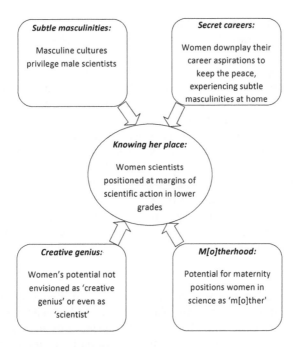

Figure 7.1 Bevan and Gatrell Framework: Knowing her place – positioning women in science

Subtle masculinities

To summarize, we argue that the subtle masculinities that pervade science perpetuate the masculine culture in science, as a result of which certain privileged and valued positions are 'reserved' for men (Miller, 1986; Puwar, 2004), while the place of women scientists is at the margins of influential activity. We note that men supported men rather than women and excluded them from decision making. Men also expected women to undertake the relational and supporting roles in laboratories, which they did, some even enjoying it, but these responsibilities did not gain respect from colleagues, and the women were not recognized as the creative scientists they wanted to be.

Secret careers

In addition, we suggest that women in heterosexual couples downplay their career aspirations whilst at home, experiencing subtle masculinities in action, enacted, albeit subtly, by their husbands and partners. The women in our study took on the bulk of the responsibilities for children and the home and 'talked up' the contribution of the male partners perhaps to uphold the notion of masculine exclusivity, even when women were the major breadwinners.

Creative genius

We further conclude that women are excluded from the of concept 'creative genius' and are sidelined into or 'placed' within supporting or relational roles on the margins of science, which do not lead to prestigious research careers. We note that conditioning of women for a life of low expectation and aspiration in science not only takes place when women start work. This means that women effectively 'career' through swerving professional paths, prevented from having a planned career path by the many structural and professional barriers erected in their way. Despite having supportive benefactors on occasions, senior scientist and medical bosses often controlled the career course of their female staff so that career paths were rarely under the individual's control.

Motherhood

Finally, we observe how women's potential for motherhood renders them other (or m[o]ther), once again placing them on the sidelines of science as they struggle both with the practical challenges of managing a home and career as well as experiencing unfair assumptions about their career orientation.

We have suggested that these four mechanisms: subtle masculinities, secret careers, creative genius and m[o]therhood, combine to preserve and strengthen what may seem subtle, but in practice are very powerful social structures that keep women firmly in their place in the lower echelons of science.

FURTHER RESEARCH AND OTHER RECOMMENDATIONS

Having established the strength of these structures, we are then faced with further questions: given that so few women scientists become leading researchers, what might need to be considered in future research and policy agendas for this situation to improve? Taking into consideration the above four contributory mechanisms, we suggest that for improvements in women's situation to happen, the implementation and measurement of policy will be insufficient: changes in approach are required if there is to be a move from the masculinized science. Organizational culture itself needs to change, and importantly this has to involve men at the top of organizations. Turning policy into practice needs to be a priority.

One area future research agendas could focus is on the lack of a feminist discourse in science. A lack of such discourse was apparent for the women interviewees and although it may be hoped that initiatives like Athena SWAN might in future encourage such debates, there is at present an omission. However, as the remit of Athena SWAN has expanded to other areas, not just science, there may be less focus on science. Future research could investigate ways in which a new discourse might to be developed and this should include both women and men. Arguably, without involving men in an alternative discourse that forces them to confront the difficulties women face, subtle masculinities in science will continue. It is noticeable that gender is not part of the science curriculum in universities, as Miller et al. (2002) note with regard to MBA courses. Although these authors comment that it shouldn't be taken as a model of practice (p. 28), some success was achieved on one course by incorporating a session on men and masculinity, and the men in particular said they had begun to think differently. We suggest that innovative ways of challenging the accepted masculine dominance in science organizations need to be investigated.

A further area for research could be an investigation of how enhanced mentoring of young women in school, in higher education and in public sector institutes might be achieved. Acknowledging that government initiatives such as Athena SWAN and its international equivalents seek to address inequities in science, nevertheless it appears from our findings that young women in science are disadvantaged at an early stage through being given inappropriate guidance – perhaps because advisors cannot imagine the scenario where young women might aspire to leading research roles in science. As an addition to the Athena initiative, we suggest that the range of the remit of Athena SWAN could be extended to include public sector institutes (other than academic institutes) where R&D is conducted but where it may only be part of a more extensive remit for such establishments.

It is important that public sector boards, CEOs and other leaders in science, both medical and non-medical, should initiate or support initiatives to change culture in their institutes to ensure that women are not afraid of articulating their concerns and needs. How can the culture change so that women are facilitated and supported in becoming leaders and decision makers in science? Culture can only change in organizations if leaders are prepared to turn the rhetoric of equality and diversity policies into reality and work hard themselves to encourage change. In the current climate of financial pressures, the needs of women scientists are often regarded as low priority. Establishing an environment where potential in women to develop can be identified should be a priority and providing a culture where they can speak openly is vital.

In addition, encouraging training schemes in public sector organizations with a focus on women returning to work after a break such as maternity leave should be the norm, not the exception. Research on why women leave science careers and what can be done to encourage women to stay in science is also called for.

Science is (erroneously) understood by most scientists and much of society to be objective and scientists, particularly male scientists, tend to see their recruitment practices as being similarly objective and don't see a need to change. We suggest that recruitment in public sector institutes should meet the highest standards so that unconscious bias training, for instance, would be included as essential. Recruitment guidance and unconscious bias training should include an understanding of how publication rates, for instance, are affected by maternity leave and childcare. One way of checking this aspect of appropriate competence of applicants could be to request their publication record over a two-year period of the applicants' choosing or requesting references to key publications only. As the HoCSTC report (2014) received in evidence, the importance of relational skills should be emphasized in healthcare science posts, particularly where research teams are to be managed. Guidance should also ensure that merit and competence is established by unbiased assessment and not based on perceived confidence alone rooted in stereotypical expectations.

In addition, following the proposed exit from the EU following the UK referendum in June 2016, Britain should not retract from the equality and diversity improvements made in the last few years and we recommend that, as a minimum, government introduces procedures to ensure that shared parental leave nationwide becomes accepted policy. To assist in this delivery, further action should be taken by government to introduce a period of leave exclusively for fathers (but without disadvantaging mothers), as certain countries (Norway, Sweden and Iceland) have already done. Meanwhile, public sector institutes should promote shared parental leave as the norm and encourage its uptake.

AND FINALLY

Finally, we recognize that the placing of women in roles that offer limited opportunities for leadership (especially in relation to research development and innovation) is a problem that extends beyond science. We have observed, for example, how the most prestigious roles within many professions (academia, law, medicine) are often, still, held by men. We suggest that the interrelated four factors described above in Figure 7.1 have potential to shed light on women's status beyond the arena of science. We hope that our framework may be used more widely in research on gender and work to shed light on studies of women's position in other fields.

Through articulating how women's place is defined, and showing how social and organizational structures combine to keep women in their 'place', we hope to contribute to research and policy discourses that challenge the status quo and seek to enhance women's place at work.

References

Academy for Healthcare Science (2016), website accessed 3 September 2016 at http://www.ahcs.ac.uk/.

ACAS (2015), 'Shared parental leave and pay', accessed 9 September 2016 at http://www.acas.org.uk/index.aspx?articleid=4911.

Acker, J. (1990), 'Hierarchies, jobs, bodies: A theory of gendered organizations', *Gender & Society*, **4**(2), 139–58.

Acker, J. (1992), 'Gendering organizational theory', in A. Mills and P. Tancred-Sherriff (eds), *Gendering Organizational Analysis*, London: Sage, 248–60.

Acker, J. (1998), 'The future of "gender and organizations": Connections and boundaries', *Gender, Work & Organization*, **5**(4), 195–206.

Adler, N. (1993), 'Competitive frontiers: Women managers in the Triad', *International Journal of Management and Organization*, **23**(2), 3–23.

Allen, S. and A. Hawkins (1999), 'Maternal gatekeeping: Mothers' beliefs and behaviours that inhibit greater father involvement in family work', *Journal of Marriage and the Family*, **61**(1), 199–212.

Annandale, E. and J. Clark (1996), 'What is gender? Feminist theory and the sociology of human reproduction', *Sociology of Health and Illness*, **18**(1), 17–44.

Archer, M. (2000), *Being Human: The Problem of Agency*, Cambridge, UK: Cambridge University Press.

Ashcraft, K.L. (1999), 'Managing maternity leave: A qualitative analysis of temporary executive succession', *Administrative Science Quarterly*, **44**(2), 240–80.

Association for Women in Science (AWIS) (2016), website accessed 3 September 2016 at http://www.awis.org/.

Athena Forum (2016), website accessed 8 September 2016 at http://www.athenaforum.org.uk.

Atkinson, R. (2001), 'The life history interview', in J. Gubrium and J. Holstein (eds), *Handbook of Interview Research: Context and Method*, Thousand Oaks, CA: Sage, 121–40.

Babcock, L. and S. Laschever (2003), *Women Don't Ask: Negotiation and the Gender Divide*, Princeton, NJ: Princeton University Press.

Babcock, L., S. Laschever, M. Gelfand and D. Small (2003), ' "Nice girls don't ask": Women negotiate less than men – and everyone pays the price', *Harvard Business Review*, **81**(10), 14–16.

Bagilhole, B., A. Powell, S. Barnard and A. Dainty (2008), *Researching Cultures in Science, Engineering and Technology: An Analysis of Current and Past Literature, Research Report Series for UKRC No. 7*, July, Loughborough, UK: UK Resource Centre for Women in Science, Department of Social Sciences, Loughborough University

Bailyn, L. (2004), 'Time in careers – careers in time', *Human Relations*, **57**(12), 1507–21.

Barr, J. and L. Birke (1998), *Common Science? Women, Science, and Knowledge*, Indianapolis, IN: Indiana University Press.

Battersby, C. (1989), *Gender and Genius: Towards a Feminist Aesthetics*, London: The Women's Press.

Battersby, C. (1998), *The Phenomenal Woman: Feminist Metaphysics and the Patterns of Identity*, Cambridge and Oxford, UK: Polity Press in association with Blackwell Publishers.

Baumgart, A. (1985), 'Women, nursing and feminism: An interview with Alice J. Baumgart by Margaret Allen', *Canadian Nurse*, January, 20–22.

BBC News (2014, 4 February), 'Microsoft's Nadella sorry for women's pay comments', accessed 8 September 2016 at http://www.bbc.co.uk/news/business-29571754.

Beck, U. and E. Beck-Gernsheim (1995), *The Normal Chaos of Love*, trans. M. Ritter and J. Wiebel, Cambridge, UK: Polity Press.

Bell, V., D. Bishop and A. Przbylski (2015), 'The debate over digital technology and young people', *BMJ Editorial*, 12 August 2015.

Benda, J. (1946 [1992]), *Le rapport d'Uriel*, Paris: Flammarion.

Bendl, R. (2008), 'Gender subtexts – reproduction of exclusion in organizational discourse', *British Journal of Management*, **19**(S1), S50–S64.

Benokraitis, N. and J. Feagin (1995), *Modern Sexism: Blatant, Subtle, and Covert Discrimination*, 2nd edition, Englewood Cliffs, NJ: Prentice Hall.

Benschop, Y. and M. Brouns (2003), 'Crumbling ivory towers: Academic organizing and its gender effects', *Gender, Work & Organization*, **10**(2), 194–212.

Bevan, V. (2009), 'Positioning women in science: Knowing her place', PhD thesis in Management Learning and Leadership, Lancaster University, UK.

Bevan, V. and M. Learmonth (2013), '"I wouldn't say it's sexism except that it's all these little subtle things": Healthcare scientists accounts of gender in healthcare laboratories', *Social Studies of Science* **43**(1), 136–58.

Bilimoria, D. and X. Liang (2012), *Gender Equity in Science and Engineering: Advancing Change in Higher Education*, New York: Routledge.

Bittman, M. and J. Pixley (1997), *The Double Life of the Family*, Sydney: Allen and Unwin.

Blackstock, C. (2004), 'Fellows keep Greenfield off Royal Society list', *The Guardian*, April 29.

Blair-Loy, M. (2003), *Competing Devotions: Career and Family Among Women Executives*, Cambridge, MA: Harvard University Press.

Blau, F.D. and M.K. Lawrence (2000), 'Gender differences in pay', *Journal of Economic Perspectives*, **14**(4), 75–99.

Blau, F.D., M.A. Ferber and A.E. Winkler (2014), *The Economics of Women, Men, and Work*, Upper Saddle River, NJ: Pearson Education, Inc.

Bowles, H., L. Babcock and L. Lai (2005), 'It depends who is asking and who you ask: Social incentives for sex differences in the propensity to initiate negotiation', *Faculty Research Working Paper Series*, Cambridge, MA: Harvard University

Bradshaw, P. (1996), 'Women as constituent directors: Re-reading current texts using a feminist-postmodernist approach', in D.M. Boje, R.P. Gephart and T.J. Thatchenkery (eds), *Postmodern Management and Organization Theory*, Thousand Oaks, CA: Sage, 95–124.

Brannen, J. and P. Moss (1991), *Managing Mothers: Dual Earner Households After Matrimony*, London: Unwin Hyman Ltd.

Brewis, J. (1999), 'How does it feel? Women managers, embodiment and changing public sector cultures', in S. Whitehead and R. Moodley (eds), *Transforming Management: Gendering Change in the Public Sector*, London: UCL Press, 84–106.

Budig, M.J. and P. England (2001), 'The wage penalty for motherhood', *American Sociological Review*, 204–25.

Burke, R. (2002), 'Career development of managerial women', in R. Burke and D. Nelson (eds), *Advancing Women's Careers*, Oxford: Blackwell, 139–60.

Burnett, S.B., C.J. Gatrell, C. Cooper and P. Sparrow (2013), 'Fathers at work: A ghost in the organizational machine', *Gender, Work & Organization*, **20**(6), 632–46.

Burris, B. (1996), 'Technocracy, patriarchy and management', in D. Collinson and J. Hearn (eds), *Men as Managers, Managers as Men: Critical Perspectives on Men, Masculinities and Managements*, London: Sage, 61–77.

Business Innovation and Skills (BIS) (2012), *HM Government – Consultation on Modern Workplaces: The Parental Leave (EU Directive) Regulations 2013 – Impact Assessment*, URN 12/1285, London: BIS.

Calás, M. and L. Smircich (1996), 'From the woman's point of view: Feminist approaches to organization studies', in S. Clegg, C. Hardy and W. Nord (eds), *Handbook of Organization Studies*, London: Sage, 218–57.

Carlson, J. and M. Crawford (2011), 'Perceptions of relational practices in the workplace', *Gender, Work & Organization*, **18**(4), 359–76.

Chandler, J. (1991), *Women Without Husbands*, Basingstoke, UK: Macmillan Education.

Charles, N. (1993), *Gender Divisions and Social Change*, Hemel Hempstead, UK: Harvester Wheatsheaf.

Charles, N. and M. Kerr (1988), *Women, Food and Families*, Manchester, UK: Manchester University Press.

Chivers, T. (2014), 'Susan Greenfield: "I'm not scaremongering"', *The Telegraph*, 24 August, accessed 9 September 2016 at http://www.tele graph.co.uk/culture/books/11050871/Susan-Greenfield-Im-not-scare mongering.html.

Church of England (2016), website accessed 9 September 2016 at https:// www.churchofengland.org/.

Citron, M.J. (1986), 'Women and the lied, 1775–1850', in J. Bowers and J. Tick (eds), *Women Making Music: The Western Art Tradition, 1150– 1950*, Urbana and Chicago: University of Illinois Press, 224–48.

Cockburn, C. (1991), *In the Way of Women: Men's Resistance to Sex Equality in Organizations*, London: Macmillan.

Code, L. (1991), *What Can She Know? Feminist Theory and the Construction of Knowledge*, Ithaca, NY and London: Cornell University Press.

Coffey, A. (1999), *The Ethnographic Self: Fieldwork and the Representation of Identity*, London: Sage.

Coles, T. (2008), 'Finding space in the field of masculinity: Lived experiences of men's masculinities', *Journal of Sociology*, **44**(3), 233–48.

Collinson, D. and J. Hearn (1994), 'Naming men as men: Implications for work, organization and management', *Gender, Work & Organization*, **1**(1), 2–22.

Collinson, D. and J. Hearn (1996), 'Breaking the silence: On men, masculinities and managements', in D. Collinson and J. Hearn (eds), *Men as Managers, Managers as Men: Critical Perspectives on Men, Masculinities and Managements*, London: Sage, 1–24.

Coltrane, S. (1996), *Family Man*, New York: Oxford University Press.

Connell, R.W. (2002), *Gender*, Cambridge, UK: Polity Press.

Connell, R.W. (2005), *Masculinities*, Berkeley, CA: University of California Press.

Cooper, C. (1992), 'The non and nom of accounting for (m)other nature', *Accounting, Auditing & Accountability Journal*, **5**(3), 16–51.

Council of Healthcare Science (2016), website accessed 9 September 2016 at http://www.councilofhealthcarescience.ac.uk/.

Creager, A.N., E. Lunbeck and L.L. Schiebinger (2001), *Feminism in Twentieth-Century Science, Technology, and Medicine*, Chicago, IL: University of Chicago Press.

Creese, M. (1998), *Ladies in the Laboratory: American and British Women in Science 1800–1900: A Survey of Their Contributions to Research*, Lanham, MD: Scarecrow Press.

Crompton, R. (1997), *Women and Work in Modern Britain*, Oxford: Oxford University Press.

Cronin, H. (1991), *The Ant and the Peacock*, Cambridge, UK: Cambridge University Press.

Cross, S. and B. Bagilhole (2002), 'Girls' jobs for the boys? Men, masculinity and non-traditional occupations', *Gender, Work & Organizations*, **9**(2), 204–26.

Cullen, D. (1994), 'Feminism, management and self-actualization', *Gender, Work & Organization*, **1**(3), 127–37.

Cunningham, J. and T. Macan (2007), 'Effects of applicant pregnancy on hiring decisions and interview ratings', *Sex Roles*, **57**(7–8), 497–508.

Davidson, M. and C. Cooper (1992), *Shattering the Glass Ceiling*, London: Paul Chapman Publishing.

Davies, A. and R. Thomas (2002), 'Gender and gendering in public service organizations: Changing professional identities under new public management', *Public Management Review*, **4**(4), 461–84.

Davies, C. (1983), 'Professionals in Bureaucracies: The Conflicts Revisited' in R. Dingwall and P. Lewis (eds), *The Sociology of the Professions: Lawyers, Doctors and Others*, London: Macmillan, 177–94.

Davies, C. (1995), *Gender and the Professional Predicament in Nursing*, Buckingham, UK: Open University Press.

Davies, C. and J. Rosser (1986), *Processes of Discrimination. A Study of Women in the NHS*, London: Department of Health.

Davis, K. (2001), ' "Peripheral and subversive": Women making connections and challenging the boundaries of the science community', *Science Education*, **85**(4), 368–409.

Davis, S., F. Neathey, J. Regan and R. Willison (2005), 'Pregnancy discrimination at work: A qualitative study', *EOC Working Paper Series*, Institute for Employment Studies, Equal Opportunities Commission.

De Beauvoir, S. ([1949] 1997), in H.M. Parshley (trans. and ed.), *The Second Sex*, London: Vintage.

De Cheveigné, S (2009), 'The career paths of women (and men) in French research', *Social Studies of Science*, **39**(1), 113–36.

Delamont, S. (2001a), *Changing Women, Unchanged Men: Sociological Perspectives on Gender in a Post-industrial Society*, Buckingham, UK: Open University Press.

Delamont, S. (2001b), 'Reflections on social exclusion', *International Studies in Sociology of Education*, **11**(1), 25–40.

Delamont, S. (2003), *Feminist Sociology*, London: Sage.

Delphy, C. and D. Leonard (1992), *Familiar Exploitation: A New Analysis of Marriage in Contemporary Western Societies*, Cambridge and Oxford, UK: Polity Press in association with Blackwell Publishers.

Dermott, E. (2014), *Intimate Fatherhood: A Sociological Analysis*, Abingdon, UK: Routledge.

DfE (2015), 'A level and other level 3 results in England 2014/2015 (provisional)', DfE, SFR 38/2015, issued 15 October, accessed 9 September 2016 at https://www.gov.uk/government/uploads/system/uploads/atta chment_data/file/467825/SFR38_2015__A_level_and_other_level_3_re sults_in_England.pdf.

Duff, W. (1807), *Letters on the Intellectual and Moral Character of Women*, Aberdeen: J. Chalmers.

Dyhouse, C. (2006), *Students: A Gendered History*, London: Routledge.

Eagly, A.H. and L.L. Carli (2007), *Through the Labyrinth: The Truth About How Women Become Leaders*, Boston, MA: Harvard Business School Press.

Eagly, A.H. and S. Karau (2002), 'Role congruity theory of prejudice toward female leaders', *Psychological Review*, **109**(3), 573–98.

Eagly, A.H., M.C. Johannesen-Schmidt and M.L. van Engen (2003), 'Transformational, transactional, and laissez-faire leadership styles: A meta-analysis comparing women and men', *Psychological Bulletin*, **129**(4), 569–91.

Edgell, S. (1980), *Middle Class Couples: A Study of Segregation, Domination and Inequality in Marriage*, London: George Allen and Unwin.

Edwards, P. and J. Wajcman (2005), *The Politics of Working Life*, Oxford: Oxford University Press.

Ehrenreich, B. and D. English (1973), *Witches, Midwives and Nurses*, New York: The Feminist Press at the City University of New York.

Ehrenreich, B. and D. English (1978), *For Her Own Good: 150 Years of the Experts' Advice to Women*, London: Pluto Press.

Ehrenreich, B. and A.R. Hochschild (2003), *Global Woman: Nannies, Maids, and Sex Workers in the New Economy*, London: Granta Books.

Eichler, M. (1980). *The Double Standard: A Feminist Critique of Feminist Social Science*, London: Croom Helm.

Eisenstein, H. (1984), *Contemporary Feminist Thought*, London: George Allen & Unwin.

Elves, M.W. and I. Gibson (2013), 'What is the scientist's role in society and how do we teach it?' *The Guardian*, 4 November, accessed 9 September

2016 at https://www.theguardian.com/higher-education-network/blog/2013/nov/04/science-in-society-policy-research.

Equalities Review (2007), *Fairness and Freedom: The Final Report of the Equalities Review*, accessed 9 September 2016 at http://webarchive.nationalarchives.gov.uk/20100807034701/http:/archive.cabinetoffice.gov.uk/equalitiesreview/upload/assets/www.theequalitiesreview.org.uk/equality_review.pdf.

Equality Act (2010), London: Office of Public Sector Information.

Equality and Human Rights Commission (EHRC) (2009), 'Equal pay strategy and position paper', accessed 9 September 2016 at https://www.equalityhumanrights.com/en/parliamentary-library/equal-pay-strategy-and-position-paper-march-2009.

Equality and Human Rights Commission (EHRC) (2016a), *Pregnancy and Maternity-related Discrimination and Disadvantage: Summary of Key Findings*, accessed 9 September 2016 at https://www.equalityhumanrights.com/en/managing-pregnancy-and-maternity-workplace/pregnancy-and-maternity-discrimination-research-findings.

Equality and Human Rights Commission (EHRC) (2016b), *Estimating the Financial Costs of Pregnancy and Maternity-related Discrimination*, accessed 31 October 2016 at https://www.equalityhumanrights.com/sites/default/files/research-report-105-cost-of-pregnancy-maternity-discrimination.pdf.

Equality Challenge Unit (ECU) (undated), 'About ECU's Athena SWAN Charter', accessed 2 June 2017 at http://www.ecu.ac.uk/equality-charters/athena-swan/about-athena-swan/.

Equality Challenge Unit (ECU) (2016), 'Athena SWAN Charter', accessed 9 September 2016 at http://www.ecu.ac.uk/equality-charters/athena-swan/.

Etzioni, A. (1964), *Modern Organizations*, Upper Saddle River, NJ: Prentice Hall.

Etzioni, A. (1969), *The Semi-professions and their Organization*, New York: Free Press.

European Council Directive (2010), 'Council Directive 2010/18/EU of 8 March 2010 implementing the revised Framework Agreement on parental leave concluded by BUSINESSEUROPE, UEAPME, CEEP and ETUC and repealing Directive 96/34/EC', *Official Journal of the European Union*, accessed 22 June 2017 at http://eur-lex.europa.eu/legal-content/EN/TXT/?uri=CELEX%3A32010L0018.

Evans, E. and C. Grant (2008), *Mama PhD*, Piscataway, NJ: Rutgers University Press.

Evetts, J. (2000), 'Analysing change in women's careers: Culture, structure and action dimensions', *Gender, Work & Organization*, 7(1), 57–67.

Fara, P. (2004), *Pandora's Breeches: Women, Science and Power in the Enlightenment*, London: Pimlico.

Featherstone, B. (2009), *Contemporary Fathering: Theory, Policy and Practice*, Bristol: Policy Press.

Fee, E. (1981), 'Is feminism a threat to scientific objectivity?' *International Journal of Women's Studies*, **4**(4), 378–92.

Fee, E. (1983), 'Women's nature and scientific objectivity', in M. Lowe and R. Hubbard (eds), *Woman's Nature: Rationalizations of Inequality*, Oxford: Pergamon Press, 9–28.

Fletcher, J. (2001), *Disappearing Acts: Gender Power and Relational Practice at Work*, Cambridge, MA and London: MIT Press.

Flicker, E. (2003), 'Between brains and breasts – women scientists in fiction film: On the marginalization and sexualization of scientific competence', *Public Understanding of Science*, **12**(3), 307–18.

Ford, J. and N. Harding (2010), ' "Get back into that kitchen, woman": Management conferences and the making of the female professional worker', *Gender, Work & Organization*, **17**(5), 503–20.

Fotaki, M. (2011), 'The sublime object of desire (for knowledge): Sexuality at work in business and management schools in England', *British Journal of Management*, **22**(1), 42–53.

Fotaki, M. (2013), 'No woman is like a man (in academia): The masculine symbolic order and the unwanted female body', *Organization Studies*, **34**(9), 1251–75.

Fotaki, M. and N. Harding (2012), 'Lacan and sexual difference in organization and management theory: Towards a hysterical academy?' *Organization*, **20**(2), 153–72.

Founier, V. and M. Keleman (2001), 'The crafting of community: Recoupling discourses of management and womanhood', *Gender Work & Organizations*, **8**(3), 267–90.

Franks, L. (2011), 'Interview: Susan Greenfield', *The Jewish Chronicle*, 24 November, accessed 26 September 2016 at http://www.thejc.com/lifestyle/lifestyle-features/58912/interview-susan-greenfield.

Freud, D. (2007), *Reducing Dependency, Increasing Opportunity: Options for the Future of Welfare to Work: An Independent Report to the Department for Work and Pensions*, accessed 9 September 2016 at http://webarchive.nationalarchives.gov.uk/20130128102031/http://dwp.gov.uk/docs/welfarereview.pdf.

Gallagher, S. (1997), 'Problem-based learning: Where did it come from, what does it do, and where is it going?', *Journal for the Education of the Gifted*, **20**(4), 332–62.

Galton, F. ([1869] 1972), in C.D. Darlington (ed.), *Hereditary Genius:*

An Inquiry into its Laws and Consequences, reprinted from 2nd (1892) edition, Gloucester, MA: Peter Smith.

Gatrell, C. (2005), *Hard Labour: The Sociology of Parenthood*, Maidenhead, UK: Open University Press.

Gatrell, C. (2006a), 'A fractional commitment? Part-time work and the maternal body'. *International Journal of Human Relations*, **18**(3), 462–75.

Gatrell, C. (2006b), 'Interviewing fathers: Feminist dilemmas in field-work', *Journal of Gender Studies*, **15**(3), 237–51.

Gatrell, C. (2007), 'Whose child is it anyway?', *The Sociological Review*, **55**(2), 352–72.

Gatrell, C. (2008), *Embodying Women's Work*, Maidenhead, UK: McGraw-Hill/Open University Press.

Gatrell, C. (2011a), 'Policy and the pregnant body at work: Strategies of secrecy, silence and supra-performance', *Gender, Work & Organization*, **18**(2), 158–81.

Gatrell, C.J. (2011b), ' "I'm a bad mum": Pregnant presenteeism and poor health at work', *Social Science & Medicine*, **72**(4), 478–85.

Gatrell, C. (2013), 'Maternal body work: How women managers and professionals negotiate pregnancy and new motherhood at work', *Human Relations*, **66**(5), 621–44.

Gatrell, C. and C. Cooper (2007), '(No) cracks in the glass ceiling: Women managers and stress and the barriers to success', in D. Bilimonia and S. Piderit (eds), *The Handbook on Women in Business and Management*, Cheltenham, UK and Northampton, MA, USA: Edward Elgar Publishing, pp. 57–77.

Gatrell, C. and E. Swan (2008), *Gender and Diversity in Management: A Concise Introduction*, London: Sage.

Gatrell, C.J., S.B. Burnett, C.L. Cooper and P. Sparrow (2014), 'Parents, perceptions and belonging: Exploring flexible working among UK fathers and mothers', *British Journal of Management*, **25**(3), 473–87.

Gershuny, J. (2011), 'Time-use surveys and the measurement of national well-being', article for UK Office for National Statistics, accessed 9 September 2016 at http://webarchive.nationalarchives.gov.uk/20160105160709/http://www.ons.gov.uk/ons/rel/environmental/time-use-surveys-and-the-measurement-of-national-well-being/article-by-jonathan-gershuny/index.html.

Gershuny, J., M. Godwin and S. Jones (1994), 'The domestic labour revolution: A process of lagged adaptation?', in M. Anderson, F. Bechhofer and J. Gershuny (eds), *The Social and Political Economy of the Household*, Oxford: Oxford University Press, pp. 151–97.

Gherardi, S. (1995), *Gender, Symbolism and Organizational Cultures*, London: Sage.

Gilligan, C. (1982), *In a Different Voice*, Cambridge, MA: Harvard University Press.

Gissing, G. (1893 [1980]), *The Odd Woman*, London: Virago.

Glucksmann, M. (2005), 'Shifting boundaries and interconnections: Extending the "total social organisation of labour"', *The Sociological Review*, **53**(S2), 19–36.

Goffman, E. (1963), *Behavior in Public Places: Notes on the Social Organization of Gatherings*, New York: Free Press.

Greenfield, S. (2002a), *SET FAIR: A Report on Women in Science, Engineering, and Technology*, accessed 9 September 2016 at www.amit-es.org/sites/default/files/pdf/publicaciones/greenfield_2003.pdf.

Greenfield, S. (2002b), 'The wrong chemistry', *The Guardian*, 28 November, accessed 4 December 2016 at https://www.theguardian.com/world/2002/nov/28/gender.uk.

Greenfield, S. (2014), *Mind Change: How Digital Technologies Are Leaving Their Mark on Our Brains*, London: Rider.

Greenfield, S. (2015), 'Championing the cause for women in science', accessed 9 September 2016 at http://www.susangreenfield.com/science/women-in-science/.

Greer, G. (1970), *The Female Eunuch*, London: MacGibbon and Kee.

Greer, G. (1991), *The Change: Women, Ageing and the Menopause*, London: Hamish.

Gregory, A. and S. Milner (2009), Editorial: 'Work–life balance: A matter of choice?' *Gender, Work & Organization*, **16**(1), 1–13.

Gregory, A. and S. Milner (2011), 'What is "new" about fatherhood?: The social construction of fatherhood in France and the UK', *Men and Masculinities*, **14**(5), 588–606.

Gregson, N. and M. Lowe (1994), *Servicing the Middle Classes: Class, Gender and Waged Domestic Labour in Contemporary Britain*, London: Routledge.

Guardian, The (2014, 12 August), 'Maternity and paternity rights', accessed 9 September 2016 at https://www.theguardian.com/money/2014/aug/12/managers-avoid-hiring-younger-women-maternity-leave.

Gueutal, H.G. and E.M. Taylor (1991), 'Employee pregnancy: The impact on organizations, pregnant employees and co-workers', *Journal of Business and Psychology*, **5**(4), 459–76.

Hacking, I. (1998), 'On being more literal about constructionism', in I. Velody and R. Williams (eds), *The Politics of Constructionsism*, London: Sage, 49–68.

Hakim, C. (1995), 'Five feminist myths about women's employment', *British Journal of Sociology*, **46**(3), 429–55.

Hakim, C. (2000), *Work-Lifestyle Choices in the 21st Century: Preference Theory*, New York: Oxford University Press.

Hakim, C. (2010), *Feminist Myths and Magic Medicine*, London: Centre for Policy Studies.

Hakim, C. (2011), 'Feminist myths and magic medicine: The flawed thinking behind calls for further equality legislation', discussion paper, London: Centre for Policy Studies.

Halpert, J.A., M.L. Wilson and J.L. Hickman (1993), 'Pregnancy as a source of bias in performance appraisals', *Journal of Organizational Behavior*, **14**(7), 649–63.

Hamilton, E. (2006), 'Whose story is it anyway? Narrative accounts of the role of women in founding and establishing family businesses', *International Small Business Journal*, **24**(3), 253–71.

Harding, S. (1991), *Whose Science? Whose Knowledge?*, Ithaca, NY: Cornell University Press.

Hart, R.A., M. Moro and J.E. Roberts (2012), 'Date of birth, family background, and the 11 plus exam: Short- and long-term consequences of the 1944 secondary education reforms in England and Wales', *Stirling Economics Discussion Paper No. 2012-10*, accessed 4 September 2016 at https://www.stir.ac.uk/media/schools/management/documents/workingpapers/SEDP-2012-10-Hart-Moro-Roberts.pdf.

Hartmann, H. (1979), 'Capitalism, patriarchy and job segregation by sex', in Z. Eisenstein (ed.), *Capitalist Patriarchy and the Case for Socialist Feminism*, New York: Monthly Review Press, 206–47.

Hartsock, N. (1983), 'The feminist standpoint: Developing a ground for a specifically feminist historical materialism', in S. Harding and M. Hintikka (eds), *Feminist Perspectives on Epistemology, Metaphysics, Methodology and Philosophy of Science*, Dordrecht: Reidel, 283–310.

Harvey, D. (1993), 'From space to place and back again: Reflections on the condition of postmodernity', in J. Bird, B. Curtis and T. Putnam et al. (eds), *Mapping the Futures. Local Cultures, Global Change*, London: Routledge, 3–29.

Haste, H. (1993), *The Sexual Metaphor*, Hemel Hempstead, UK: Harvester Wheatsheaf.

Hawkins, A. and D. Dollahite (eds) (1997), 'Beyond the role inadequacy perspective', in *Generative Fathering: Beyond Deficit Perspectives*, Thousand Oaks, CA: Sage, 3–16.

Haynes, K. (2006), 'Insider account: A therapeutic journey? Reflections on the effects of research on the researcher and participants', *Qualitative*

Research in Organizations and Management: An International Journal, **1**(3), 204–21.

Haynes, K. (2008), 'Transforming identities: Accounting professionals and the transition to motherhood', *Critical Perspectives on Accounting*, **19**(5), 620–42.

Health and Care Professions Council (HCPC) (2016), website accessed 3 September 2016 at http://www.hcpc-uk.co.uk/.

Hearn, J. (1982), 'Notes on patriarchy professionalism and the semi-professions', *Sociology*, **16**(2), 184–202.

Hearn, J. (1987), *The Gender of Oppression: Men, Masculinity, and the Critique of Marxism*, Brighton, UK: Harvester Wheatsheaf.

Hearn, J. (1998), 'Theorizing men and men's theorizing: Men's discursive practices in theorizing men', *Theory and Society*, **27**(6), 781–816.

Hearn, J. (2002), 'Alternative conceptualizations and theoretical perspectives in identities and organizational culture: A personal review of research on men in organizations', in I. Aaltio and A. Mills (eds), *Gender, Identity and the Culture of Organizations*, London: Routledge, 39–56.

Hearn J., R. Piekkari and M. Jyrkinen (2009), *Managers Talk About Gender: What Managers in Large Transnational Corporations Say About Gender Policies, Structure and Practices*, Helsinki: Hanken School of Economics Research Reports.

Hersby, M.D., M.K. Ryan and J. Jetten (2009), 'Getting together to get ahead: The impact of social structure on women's networking', *British Journal of Management*, **20**(4), 415–30.

Hewlett, S.-A., C. Buck Luce and L.J. Servon et al. (2008), *The Athena Factor: Reversing the Brain Drain in Science, Engineering and Technology*, Boston, MA: Harvard Business School Publishing.

Hill, C., C. Corbett and A. St Rose (2010), *Why So Few? Women in Science, Technology, Engineering, and Mathematics*, Washington: American Association of University Women.

Hochschild, A.R. (2003), *The Second Shift*, New York: Penguin Books.

Hollenshead, C. (2003), 'Women in the academy: Confronting barriers to equality', in L.S. Hornig (ed.), *Equal Rites, Unequal Outcomes: Women in American Research Universities*, New York: Kluwer Academic/ Plenum Publishers, 211–25.

Hollway, W. (1996), 'Masters and men in the transition from factory hands to sentimental workers', in D. Collinson and J. Hearn (eds), *Men as Managers, Managers as Men: Critical Perspectives on Men, Masculinities and Managements*, London: Sage, 25–42.

Holstein, J. and J. Gubrium (1997), 'Active interviewing', in D. Silverman (ed.), *Qualitative Research: Theory, Method and Practice*, London: Sage, 113–29.

Holter, Ø.G. (2007), 'Men's work and family reconciliation in Europe'. *Men and Masculinities*, **9**(4), 425–56.

Höpfl, H. (2000), 'Organizing women and organizing women's writing', *Gender, Work & Organization*, **7**(2), 98–105.

Höpfl, H. and P. Hornby Atkinson (2000), 'The future of women's careers', in A. Collin and R. Young (eds), *The Future of Career*, Cambridge, UK: Cambridge University Press, 130–43.

Hosek, S., A. Cox and B. Ghosh-Dastidar et al. (2005), *Gender Differences in Major Federal External Grant Programs*, Santa Monica, CA: RAND Corporation.

House of Commons Science and Technology Committee (HoCSTC) (2014), *Women in Scientific Careers, Sixth Report of Session 2013–2014*, accessed 2 June 2017 at https://www.publications.parliament.uk/pa/cm201314/cmselect/cmsctech/701/701.pdf.

House of Commons Women and Equalities Committee on Pregnancy and Discrimination (2016–17), *Pregnancy and Maternity Discrimination*, accessed 22 June 2017 at https://www.publications.parliament.uk/pa/cm201617/cmselect/cmwomeq/90/9002.htm.

Husu, L. (2001), 'Sexism, support and survival. Academic women and hidden discrimination in Finland', *Social Psychological Studies 6*, Helsinki. University of Helsinki.

Institute of Biomedical Science (IBMS) (2016), website accessed 9 September 2016 at https://www.ibms.org/.

James, C.G. (2007), 'Law's response to pregnancy/workplace conflicts: A critique', *Feminist Legal Studies*, **15**(2), 167–88.

Janeway, E. (1971), *Man's World, Women's Place: A Study in Social Mythology*, New York: Dell Publishing.

Jervey, G. (2005), 'She makes more money than he does', *Money*, **34**(5), 41–4.

Johnson, J. (2001), 'In depth interviewing', in J. Gubrium and J. Holstein (eds), *Handbook of Interview Research: Context and Method*, Thousand Oaks, CA: Sage, 103–20.

Kant, I. (1790 [1951]), *The Critique of Judgement*, trans. J.H. Bernard, New York: Hafner.

Kanter, R. (1977), *Men and Women of the Corporation*, New York: Basic Books.

Keller, E. (1983) *A Feeling for the Organism: The Life and Work of Barbara McClintock*, San Francisco, CA: Freeman.

Keller, E. (1985), *Reflections on Gender and Science*, New Haven, CT and London: Yale University Press.

Keller, E.F. and H.E. Longino (1996), *Feminism and Science*, Oxford: Oxford University Press.

Kemelgor, C. and H. Etzkowitz (2001), 'Overcoming isolation: Women's dilemmas in American academic science', *Minerva*, **39**(2), 153–74.

Kerckhoff, A. (1995), 'Institutional arrangements and stratification processes in industrial societies', *Annual Review of Sociology*, **21**(3), 323–47.

Kerfoot, D. and D. Knights (1993), 'Management, masculinity and manipulation: From paternalism to corporate strategy in financial services in Britain', *Journal of Management Studies*, **30**(4), 659–78.

Kimmel, M. (1993), 'The new organization man: What does he really want?', *Harvard Business Review*, November–December.

Kirkup, G., A. Zalevski, T. Maruyama and I. Batool (2010), *Women and Men in Science, Engineering and Technology: The UK Statistics Guide 2010*, Bradford, UK: UK Resource Centre for Women (UKRC).

Klug, A. (1968 [1980]), 'Rosalind Franklin and the discovery of the structure of DNA', reprinted in G. Stent (ed.) (1980), *James D. Watson: The Double Helix: A Personal Account of the Discovery of the Structure of DNA*, London: George Weidenfeld and Nicolson, 153–8.

Kumra, S. and S. Vinnicombe (2008), 'A study of the promotion to partner process in a professional services firm: How women are disadvantaged', *British Journal of Management*, **19**(S1), S65–S74.

Küskü, F., M. Özbilgin and L. Özkale (2007), 'Against the tide: Gendered prejudice and disadvantage in engineering', *Gender, Work & Organization*, **14**(2), 109–29.

Lane, N. (1999). 'Why are there so few women in science?', *Nature Debates*, 9 September, accessed 9 September 2016 at http://www.nature.com/nature/debates/women/women_contents.html.

Laws, J. (1975), 'The psychology of tokenism, *Sex Roles*, **1**(1), 51–67.

Learmonth, M. (2009), ' "Girls" working together without "teams": How to avoid the colonization of management language', *Human Relations*, **62**(12), 1887–906.

Liff, S. and K. Ward (2001), 'Distorted views through the glass ceiling: The construction of women's understanding of promotion and senior management positions', *Gender Work & Organization*, **8**(1), 19–36.

Littleton, S., M. Arthur and D. Rousseau (2000), 'The future of boundaryless careers', in A. Collin and R. Young (eds), *The Future of Career*, Cambridge, UK: Cambridge University Press, 101–15.

Longhurst, R. (2001), *Bodies: Exploring Fluid Boundaries*, London: Routledge.

Longhurst, R. (2008), *Maternities: Gender, Bodies and Space*, New York/London: Routledge.

Lupton, B. (2000), 'Maintaining masculinity: Men who do "women's work" ', *British Journal of Management*, **11**(Special Issue), S33–S48.

Lupton, D. and L. Barclay (1997), *Constructing Fatherhood: Discourses and Experiences*, London: Sage.

Lwoff, A. (1968 [1980]), 'Truth, truth, what is truth' (about how the structure of DNA was discovered)?' in G. Stent (ed.), *James D. Watson: The Double Helix: A Personal Account of the Discovery of the Structure of DNA*, London: George Weidenfeld and Nicolson, 225–34.

Lynn, R. and S. Kanazawa (2011), 'A longitudinal study of sex differences in intelligence at ages 7, 11 and 16 years', *Personality and Individual Differences*, **51**(3), 321–4.

Mair, V. (2010), 'Baroness Greenfield drops legal action against the Royal Institution', *Civil Society News*, 5 May, accessed 9 September 2016 at http://www.civilsociety.co.uk/governance/news/content/6560/baroness_greenfield_drops_sex_discrimination_case_against_the_royal_institution.

Mäkelä, L. (2005), 'Pregnancy and leader–follower dyadic relationships: A research agenda', *Equal Opportunities International*, **24**(3/4), 50–72.

Mallon, M., J. Duberley and L. Cohen (2005), 'Careers in public sector science: Orientations and implications', *R&D Management*, **35**(4), 395–407.

Mansfield, P. and J. Collard (1988), *The Beginning of the Rest of your Life?* Basingstoke, UK: Macmillan.

Maranda, M.-F. and Y. Comeau (2000), 'Some contributions of sociology to the understanding of the career', in A. Collin and R. Young (eds), *The Future of Career*, Cambridge, UK: Cambridge University Press, 37–52.

Marshall, J. (1995), *Women Managers Moving On: Exploring Career and Life Choices*, London: Routledge.

Marshall, J. (2000), 'Living lives of change: Examining facets of women managers' career stories', in M. Arthur, M. Peiperl, R. Goffee and T. Morris (eds), *Career Frontiers: New Conceptions of Working Lives*, Oxford: Oxford University Press, 202–27.

Martin, B. (1984), 'Mother wouldn't like it: House work as magic', *Theory, Culture and Society*, **2**(2), 19–36.

Martin, E. (1989), *The Woman in the Body: A Cultural Analysis of Reproduction*, Milton Keynes, UK: Open University Press.

Martin, J. (1990), 'Deconstructing organizational taboos: The suppression of gender conflict in organizations', *Organization Science*, **1**(4), 339–59.

Maushart, S. (2003), *Wifework: What Marriage Really Means For Women*, London: Bloomsbury.

McCarthy, P. and H. Spanner (2000), 'Peter McCarthy talks to Professor Susan Greenfield', *Third Way*, October, 20.

McMahon, A. (1999), *Taking Care of Men.* Cambridge: Cambridge University Press.

Mill, J.S. (1869 [2006]), 'The subjection of women', in A. Rossi (ed.) (2006), *Essays on Sex Equality*, Chicago, IL: University of Chicago Press, 125–242.

Miller, J. (1986), *Towards a New Psychology of Women*, London: Penguin.

Miller, S., R. Hagen and M. Johnson (2002), 'Divergent identities? Professions, management and gender', *Public Money and Management*, **22**(1), 25–30.

Miller, T. (2005). *Making Sense of Motherhood: A Narrative Approach*, Cambridge, UK: Cambridge University Press.

Miller, T. (2011), 'Falling back into gender? Men's narratives and practices around first-time fatherhood', *Sociology*, **45**(6), 1094–109.

Millett, K. (1970 [2000]), *Sexual Politics*, Urbana and Chicago, IL: University of Illinois Press.

Millward, L.J. (2006), 'The transition to motherhood in an organizational context: An interpretative phenomenological analysis', *Journal of Occupational and Organizational Psychology*, **79**(3), 315–33.

Monosson, E. (ed.) (2008), *Motherhood, the Elephant in the Laboratory*, Ithaca, NY: Cornell University Press.

Morrison, A., R. White and E. van Velsor (1992), *Breaking the Glass Ceiling*, Reading, MA: Addison Wesley.

Morrison, H. (1878 [2010]), *My Summer in the Kitchen*, Whitefish, MT: Kessinger Publishing.

Mullin, A. (2005), *Reconceiving Pregnancy and Childcare: Ethics, Experience, and Reproductive Labor*, New York: Cambridge University Press.

My Family Care and Women's Business Council (2016), *Shared Parental Leave: Where Are We Now?*, accessed 22 June 2017 at https://www.myfamilycare.co.uk/resources/white-papers/shared-parental-leave-where-are-we-now.

Newman, J. (1995), 'Gender and cultural change', in C. Itzin and J. Newman (eds), *Gender, Culture and Organizational Change. Putting Theory into Practice*, London and New York: Routledge, 11–29.

Newsom Report (1963), *Half Our Future*, a report from the Central Advisory Council for Education, London: Her Majesty's Stationery Office.

NHS Careers (2016), accessed 9 September 2016 at https://www.healthcareers.nhs.uk/.

Nobel Prize (2017), website, accessed 12 July 2017 at http://www.nobelprize.org/.

Noddings, N. (1984), *Caring: A Feminine Approach to Ethics and Moral Education*, Berkeley, CA: University of California Press.

Nutley, S., S. Perrott and F. Wilson (2002), Editorial: 'Gender, the professions and public management', *Public Money & Management*, **22**(1), 3–4.

Oakley, A. (1974), *The Sociology of Housework*, London: Martin Robertson.

O'Brien, M. (1981), *The Politics of Reproduction*, London: Routledge and Kegan Paul.

Office for National Statistics (ONS) (2006), *Time Use Survey 2005*, accessed 2 September 2016 at http://www.timeuse.org/sites/ctur/files/public/ctur_report/1905/lader_short_and_gershuny_2005_kight_diary.pdf.

Office for National Statistics (ONS) (2013), *Full Report – Women in the Labour Market: 2013*, accessed 2 September 2016 at http://webarchive.nationalarchives.gov.uk/20160105160709/http://www.ons.gov.uk/ons/dcp171776_328352.pdf.

Office for National Statistics (ONS) (2016), 'UK Labour market: August 2016: Estimates of employment, unemployment, economic inactivity and other employment-related statistics for the UK', accessed 2 September 2016 at http://www.ons.gov.uk/employmentandlabourmarket/peopleinwork/employmentandemployeetypes/bulletins/uklabourmarket/august2016#main-points-for-april-to-june-2016.

Organisation for Economic Co-operation and Development (OECD) (2012), *Gender Equality in Education, Employment and Entrepreneurship: Final Report to the MSM 2012*, accessed 23 June 2017 at https://www.oecd.org/employment/50423364.pdf.

Palmer, C. and K. Yandell (2013), 'Life sciences salary survey – 2013', *The Scientist*, **27**(11), 1.

Parrington, J. (1996), 'The intelligence fraud', *Socialist Review*, Issue 196, accessed 9 September 2016 at http://pubs.socialistreviewindex.org.uk/sr196/parrington.htm.

Parsons, T. (1971), 'The normal American family', in B. Adams and T. Weirath (eds), *Readings on the Sociology of the Family*, Chicago, IL: Markham, 53–6.

Peiperl, M. and M. Arthur (2000), 'Topics for conversation: Career themes old and new', in M. Peiperl, M. Arthur, R. Goffee and T. Morris (eds), *Career Frontiers: New Conceptions of Working Lives*, Oxford: Oxford University Press, 1–20.

Pennell, H. and A. West (2003), *Underachievement in Schools*, London: Routledge.

Perry, N. (1992), 'If you can't join 'em, beat 'em', *Fortune*, 21 September, 58–9.

Peterson, H. (2010), 'The gendered construction of technical self-confidence: Women's negotiated positions in male dominated, technical work settings', *International Journal of Gender, Science and Technology*, **2**(1), 65–88.

Phipps, A. (2008), *Women in Science, Engineering and Technology: Three Decades of UK Initiatives*, Stoke-on-Trent, UK: Trentham Books.

Pinker, S. (2008), *The Sexual Paradox: Men, Women and the Real Gender Gap*, London: Scribner.

Piper, A. (1998), 'Rosalind Franklin: Light on a dark lady', *Trends in Biochemical Sciences*, **23**(4), 151–4.

Posen, M., D. Templer and V. Forward et al. (2005), 'Publication rates of male and female academic clinical psychologists in California', *Psychological Reports*, **97**(3), 898–902.

Potuchek, J. (1997), *Who Supports the Family? Gender and Breadwinning in Dual-earner Marriages*, Palo Alto, CA: Stanford University Press.

Prather, J. (1971), 'Why can't women be more like men? A summary of the sociopsychological factors underlining women's advantage in the professions', in L. Fidell and J. DeLameter (eds), *Women in the Professions: What's All the Fuss About?* London: Sage, 14–24.

Pringle, R. (1998), *Sex and Medicine: Gender, Power and Authority in the Medical Profession*, Cambridge, UK: Cambridge University Press.

Puwar, N. (2004), *Space Invaders: Race, Gender and Bodies Out of Place*, Oxford and New York: Berg.

Radford, T. (2004), 'The Guardian profile: Susan Greenfield', 30 April, accessed 9 September 2016 at http://www.theguardian.com/uk/2004/apr/30/science.highereducation.

Radin, M.J. (1996), *Contested Commodities: The Trouble with Trade in Sex, Children, Body Parts and Other Things*, Cambridge, MA: Harvard University Press.

Ragins, B.R. and J.L. Cotton (1999), 'Mentor functions and outcomes: A comparison of men and women in formal and informal mentoring relationships', *Journal of Applied Psychology*, **84**(4), 520–50.

Ramsay, K. and G. Letherby (2006), 'The experience of academic non-mothers in the gendered university', *Gender, Work & Organization*, **13**(1), 25–44.

Ratcliffe, R. (2013), 'The gender gap at universities: Where are all the men?', *The Guardian*, 29 January, accessed 9 September 2016 at http://www.theguardian.com/education/datablog/2013/jan/29/how-many-men-and-women-are-studying-at-my-university#data.

Reis, S. (2003), 'Gifted girls, twenty-five years later: Hopes realized and new challenges found', *Roeper Review*, **25**(4), 154–8.

Reskin, B.F. and D.B. McBrier (2000), 'Why not ascription? Organizations' employment of male and female managers', *American Sociological Review*, **65**(2) 210–33.

Rhoton, L. (2011), 'Distancing as a gendered barrier: Understanding women scientists' gender practices', *Gender & Society*, **25**(6), 696–716.

Ribbens, J. (1994), *Mothers and their Children: A Feminist Sociology of Childrearing*, London: Sage.

Roeper, A. (1978), 'The young gifted girl', Roeper Review, **1**(1), 6–8.

Rosser, S. (2004), *The Science Glass Ceiling: Academic Women Scientists and the Struggle to Succeed*, London: Routledge.

Rosser, S.V. and M.Z. Taylor (2009), 'Why are we still worried about women in science?', *Academe*, **95**(3), 7–10.

Rossi, A. (1965), 'Women in science: Why so few?' *Science*, **148**(3674), 1196–202.

Rossiter, M. (1982), *Women Scientists in America: Struggles and Strategies to 1940*, Baltimore, MD: Johns Hopkins University Press.

Roth, W. and G. Sonnert (2011), 'The costs and benefits of red tape: Anti-bureaucratic structure and gender inequity in a science research organization', *Social Studies of Science*, **41**(3), 385–409.

Rouse, J. and N. Sappleton (2009), 'Managing maternity fairly and productively support for small employers', *International Small Business Journal*, **27**(2), 215–25.

Rousseau, J.-J. ([1758] 2009), in C. Kelly and E. Grace (eds), *Rousseau on Women, Love, and Family*, Hanover, NH: University Press of New England.

Royal Society (2017), website, accessed 10 July 2017 at https://royalsoci ety.org/.

Ruddick, S. (1980), 'Maternal thinking', *Feminist Studies*, **6**(2), 342–67.

Ruddick, S. (1989), *Maternal Thinking: Towards a Politics of Peace*, London: The Women's Press.

Rudman, L. (1998), 'Self promotion as a risk factor for women: The costs and benefits of counterstereotypical impression management', *Journal of Personality and Social Psychology*, **74**(3), 629–46.

Ryan, M.K. and S.A. Haslam (2007), 'The glass cliff: Exploring the dynamics surrounding the appointment of women to precarious leadership positions', *Academy of Management Review*, **32**(2), 549–72.

Sandberg, S. (2014), *Lean in for Graduates*, New York: Alfred A Knopf.

Sayre, A. (1975), *Rosalind Franklin and DNA*, New York: W.W. Norton & Co.

Schein, V.E. (2001), 'A global look at psychological barriers to women's progress in management', *Journal of Social Issues*, **57**(4), 675–88.

Schein, V.E. and M. Davidson (1993), 'Think manager think male', *Management Development Review*, **6**(3), 24–8.

Schiebinger, L. (1989), *The Mind Has No Sex. Women in the Origins of Modern Science*, London: Harvard University Press.

Schiebinger, L. (1999), *Has Feminism Changed Science?*, Cambridge, MA: Harvard University Press.

Seligman, M. and J. Weiss (1980), 'Coping behavior: Learned helplessness, physiological change and learned inactivity', *Behaviour Research and Therapy*, **18**(5), 459–512.

Sheppard, D. (1989), 'Organizations, power and sexuality: The image and self image of women managers', in J. Hearn, D. Sheppard, P. Tancred-Sherriff and G. Burrell (eds), *The Sexuality of Organizations*, London: Sage, 139–57.

Simonton, D. (1988), *Scientific Genius: A Psychology of Science*, Cambridge, UK: Press Syndicate of the University of Cambridge.

Sims, D., S. Fineman and Y. Gabriel (1993), *Organizing and Organizations: An Introduction*, London: Sage.

Sismondo, S. (1995), 'The scientific domains of feminist standpoints', *Perspectives on Science*, **3**(1), 49–65.

Smith, D. (1974), 'Women's perspective as a radical critique to sociology', *Sociological Inquiry*, **44**(1), 7–13.

Smith, D. (1987), *The Everyday World as Problematic: A Feminist Sociology*, Lebanon, NH: Northeastern University Press.

Sonnert, G. and G. Holton (1995), *Who Succeeds in Science? The Gender Dimension*, New York: Rutgers Press.

Stanley, L. and S. Wise (1983), *Breaking Out: Feminist Consciousness and Feminist Research*, London: Routledge and Kegan Paul.

Stead, V. and C. Elliott (2009), *Women's Leadership*, Basingstoke, UK: Palgrave Macmillan.

Stent, G. (1980), *James D. Watson: The Double Helix: A Personal Account of the Discovery of the Structure of DNA*, London: George Weidenfeld and Nicolson.

Stubbs, B. (2016), 'Student performance analysis', accessed 9 September 2016 at http://www.bstubbs.co.uk/new.htm.

Tannen, D. (1992), *You Just Don't Understand: Women and Men in Conversation*, London: Virago.

Tannen, D. (2008), 'The double bind', in S. Morrison (ed.), *Thirty Ways of Looking at Hillary*, New York: Harper Collins, 126–39.

Tracy, S.J. and K.D. Rivera (2010), 'Endorsing equity and applauding stay-at-home moms: How male voices on work-life reveal aversive sexism and flickers of transformation', *Management Communication Quarterly*, **24**(1), 3–43.

Tyler, I. (2000), 'Reframing pregnant embodiment', in S. Ahmed, J. Kilby and S. Lury et al. (eds), *Transformations: Thinking Through Feminism*, London: Routledge, 288–301.

UK Government (2015), 'Shared parental leave and pay', accessed 9 September 2016 at https://www.gov.uk/shared-parental-leave-and-pay/overview.

UK Resource Centre for Women (UKRC) (2008), accessed 9 September 2016 at http://webarchive.nationalarchives.gov.uk/20080107205445/ukrc4setwomen.org/.

Valian, V. (2000), 'The advancement of women in science and engineering', in *Women in the Chemical Workforce. A Workshop Report to the Chemical Sciences Roundtable*, Washington DC: National Academies Press, 24–37.

Valian, V. (2004), 'Beyond gender schemas: Improving the advancement of women in academia', *NWSA Journal*, **16**(1), 207–20.

Van den Brink, M., Y. Benschop and W. Jansen (2010), 'Transparency in academic recruitment: A problematic tool for gender equality?', *Organization Studies*, **31**(11), 1459–83.

Vinnicombe, S. and J. Bank (2003), *Women with Attitude: Lessons for Career Management*, London: Routledge.

Vinnicombe, S., E. Doldor and R. Sealy et al. (2015), *The Female FTSE Board Report 2015: Putting the UK Progress into a Global Perspective*, Cranfield, UK: Cranfield University School of Management.

Vogler, C. (1998) 'Money in the household: Some underlying issues of power', *The Sociological Review*, **46**(4), 687–713.

Vogler, C. and J. Pahl (1994), 'Money, power and inequality within marriage'. *The Sociological Review*, **42**(2), 263–88.

Wajcman, J. (1991), *Feminism Confronts Technology*, Cambridge, UK: Polity Press.

Wajcman, J. (1998), *Managing like a Man: Women and Men in Corporate Management*, Cambridge and Oxford: Polity Press in association with Blackwell.

Walby, S. (1989), 'Theorising patriarchy', *Sociology*, **23**(2), 213–34.

Warren, S. and J. Brewis (2004), 'Matter over mind? Examining the experience of pregnancy', *Sociology*, **38**(2), 219–36.

Watson, J. (1968 [1980]), 'The double helix', in G. Stent (ed.) (1980), *James D. Watson: The Double Helix: A Personal Account of the Discovery of the Structure of DNA*, London: George Weidenfeld and Nicolson, 1–136.

Weber, M. (1949), *The Methodology of the Social Sciences*, New York: The Free Press.

Weiss, R. (1979), *Going it Alone: The Family Life and Social Situation of the Single Parent*, New York: Basic Books.

White, B., C. Cox and C. Cooper (1992), *Women's Career Development: A Study of High Flyers*, Oxford: Blackwell.

Whyte, W. (1993), *Street Corner Society: The Social Structure of an Italian Slum*, 4th edition, Chicago, IL: University of Chicago Press.

Williams, J. (2000), *Unbending Gender: Why Family and Work Conflict and What to Do About It*, Oxford: Oxford University Press.

Witz, A. (1992), *Professions and Patriarchy*, London and New York: Routledge.

Witz, A. (2000), 'Whose body matters? Feminist sociology and the corporeal turn in sociology and feminism', *Body & Society*, **6**(2), 1–24.

Wollstonecraft, M. (1792 [1999]), *A Vindication of the Rights of Woman*, New York: Bartleby.com.

Women into Science, Engineering and Construction (WISE), website accessed 9 September 2016 at http://www.wisecampaign.org.uk/.

Women & Work Commission (WWC) (2006a), *Shaping a Fairer Future*, accessed 9 September 2016 at http://webarchive.nationalarchives.gov.uk/20100212235759/http:/www.equalities.gov.uk/pdf/297158_WWC_Report_acc.pdf.

Women & Work Commission (WWC) (2006b), *Government Action Plan: Implementing the Women and Work Commission Recommendations*, accessed 7 April 2016 at https://www.researchonline.org.uk/sds/search/download.do;jsessionid=ACFD4955DB48DD812F2E58886CE3F6E5?ref=B2421.

Women-Related Web Sites in Science/Technology (2017), accessed 2 June 2017 ay http://userpages.umbc.edu/~korenman/wmst/links_sci.html.

Xie, Y. and K. Shauman (2003), *Women in Science: Career Processes and Outcomes*, Cambridge, MA: Harvard University Press.

Zuckerman, H. (1991), 'The careers of men and women scientists: A review of current research', in H. Zuckerman, J. Cole and J. Bruer (eds), *The Outer Circle: Women in the Scientific Community*, New Haven, CT: Yale University Press, 27–56.

Appendix

Table A.1 *Summary of information of respondents: By age, marital status and public or private sector employment*

Pseudonym	Interview Order	Age (Years)			Married	Children	Public Sector	Private Sector
		<30	30–50	>50				
Angela	1	–	–	√	√	√	√	–
Bob	2	–	–	√	√	√	√	–
Carla	3	–	√	–	–	√	√	–
Diana	4	–	√	–	–	–	√	–
Essie	5	–	–	√	√	√	√	–
Ida	6	–	√	–	–	–	√	–
Jacky	7	√	–	–	–	–	√	–
Francis	8	–	√	–	√	√	–	√
George	9	–	√	–	√	√	–	√
Harry	10	–	√	–	√	√	–	√
Kate	11	√	–	–	–	–	√	–
Margo	12	–	–	√	√	√	–	√
Larry	13	–	–	√	√	√	–	√
Elena	14	–	√	–	√	√	√	–
Lisa	15	–	√	–	√	√	√	–
Maggie	16	–	–	√	√	–	√	–
Nicola	17	–	√	–	√	√	√	–
Elizabeth	18	–	√	–	√	–	√	–
Paula	19	–	–	√	√	√	√	–
Rosa	20	–	–	√	√	√	√	–
Sarah	21	–	√	–	–	–	√	–
Will	22	–	–	√	√	√	√	–
Alec	23	–	–	√	√	√	√	–
Tina	24	–	–	√	√	√	√	–
Vanessa	25	√	–	–	–	–	√	–
Lily	26	–	√	–	–	–	√	–
Carolyn	27	–	–	√	√	–	√	–
Alice	28	–	–	√	√	√	√	–
Ben	29	–	√	–	√	–	√	–
Beth	30	–	√	–	√	–	–	√
Zoe	31	√	–	–	–	–	√	–

Pseudonym	Interview Order	Age (Years)			Married	Children	Public Sector	Private Sector
		<30	30–50	>50				
Deidre	32	–	–	√	–	√	√	–
Ella	33	–	–	√	√	–	√	–
Gina	34	–	√	–	√	√	√	–
Frank	35	–	–	√	√	√	√	–
Harriet	36	–	√	–	–	√	√	–
Ian	37	√	–	–	–	–	√	–
Jane	38	–	–	√	√	√	√	–
Mary	39	–	–	√	√	√	√	–
Liam	40	–	–	√	√	–	√	–
Ralph	41	–	–	√	√	√	√	–
Nina	42	–	√	–	√	√	√	–
James	43	–	–	√	√	√	√	–
Olivia	44	√	–	–	√	–	√	–
Polly	45	√	–	–	–	–	√	–
Tim	46	–	–	√	√	√	√	–
Rachel	47	–	√	–	–	–	√	–
Totals	47	7	18	22	33	28	41	6

Table A.2 *Summary of information of respondents: By professional group,*
professional qualifications, current employment, practising
scientists and research experience

Pseudonym	Professional Group	Professional Qualifications	Position	Practising Scientist	Experienced Research Scientist
Angela	HCS	√	Lab director	√	√
Bob	HCS	√	Lab manager	√	–
Carla	HCS	√	Lab manager	√	√
Diana	HCS	√	Senior HCS	√	√
Essie	HCS	√	Corporate manager	–	√
Ida	HCS	–	HCS	√	√
Jacky	HCS	√	HCS	√	√
Francis	HCS	–	Senior HCS	√	√
George	HCS	–	Senior HCS	√	√
Harry	HCS	–	Senior HCS	√	√
Kate	HCS	√	HCS	√	–
Margo	HCS	–	Senior HCS	√	√
Larry	HCS	√	Company manager	–	√
Elena	HCS	–	HCS	√	√
Lisa	HCS	–	Prof/Senior HCS	√	√
Maggie	HCS	√	HCS	√	–
Nicola	HCS	√	HCS	√	–
Elizabeth	HCS	–	HCS	√	√
Paula	HCS	√	Senior academic	√	√
Rosa	HCS	√	HCS	√	–
Sarah	HCS	–	Senior HCS	√	√
Will	HCS	–	Senior HCS	√	√
Alec	HCS	–	Senior HCS	√	√
Tina	HCS	–	HCS	√	√
Vanessa	HCS	–	HCS	√	√
Lily	HCS	–	Senior HCS	√	√
Carolyn	HCS	√	Deputy CEO	–	√
Alice	HCS	–	Senior HCS	√	√
Ben	HCS	–	HCS	√	√
Beth	HCS	√	Company manager	–	√
Zoe	HCS	–	HCS	√	√
Deidre	Medical	√	Medical	–	√
Ella	HCS	√	Professor	√	√
Gina	HCS	–	Senior HCS	√	√
Frank	Medical	√	CEO	–	√
Harriet	HCS	–	Senior HCS	√	√
Ian	HCS	–	HCS	√	√

Pseudonym	Professional Group	Professional Qualifications	Position	Practising Scientist	Experienced Research Scientist
Jane	Medical	√	CEO	–	√
Mary	HCS	√	HCS	√	√
Liam	Medical	√	CEO	–	√
Ralph	HCS	–	CEO	–	√
Nina	HCS	–	HCS	√	√
James	HCS	√	Lab director	√	√
Olivia	HCS	–	HCS	√	√
Polly	HCS	–	HCS	√	√
Tim	HCS	√	Senior HCS	√	√
Rachel	HCS	–	HCS	√	√
Totals		22		38	42

Note: CEO = chief executive officer; HCS = healthcare science/scientist.

Table A.3 *Summary of information of respondents: Academic qualifications and mode of study*

Pseudonym	Highest Qualification	Mode of Study	No First Degree	Part-time First Degree	Part-time Masters	Part-time PhD
Angela	PhD	P/T	–	–	√	√
Bob	BSc	P/T	–	√	–	–
Carla	MSc	P/T	√	–	√	–
Diana	PhD	P/T	√	–	√	√
Essie	MBA	P/T	√	–	√	–
Ida	PhD	P/T	–	–	√	√
Jacky	MSc	P/T	–	√	√	–
Francis	PhD	F/T	–	–	–	–
George	PhD	F/T	–	–	–	–
Harry	PhD	F/T	–	–	–	–
Kate	BSc	F/T	–	–	–	–
Margo	PhD	P/T	–	–	√	√
Larry	MSc	P/T	√	–	√	–
Elena	PhD	F/T	–	–	–	–
Lisa	PhD	F/T	–	–	–	–
Maggie	MSc	P/T	√	–	√	–
Nicola	MSc	P/T	√	–	√	–
Elizabeth	PhD	F/T	–	–	–	–
Paula	PhD	P/T	√	–	√	√
Rosa	MSc	P/T	√	–	√	–
Sarah	PhD	P/T	–	√	√	√
Will	PhD	F/T	–	–	–	–
Alec	PhD	P/T	–	–	√	√
Tina	MSc	P/T	–	√	√	–
Vanessa	PhD	P/T	–	–	–	√
Lily	MSc	P/T	–	–	√	–
Carolyn	PhD	F/T	–	–	–	–
Alice	PhD	F/T	–	–	–	–
Ben	PhD	F/T	–	–	–	–
Beth	PhD	P/T	–	√	√	√
Zoe	BSc	F/T	–	–	–	–
Deidre	Medical	F/T	–	–	–	–
Ella	PhD	P/T	√	–	√	√
Gina	PhD	F/T	–	–	–	–
Frank	Medical	F/T	–	–	–	–
Harriet	PhD	P/T	–	–	√	√
Ian	BSc	F/T	–	–	–	–
Jane	Medical	F/T	–	–	–	–
Mary	MSc	P/T	√	√	√	–
Liam	Medical	F/T	–	–	–	–
Ralph	PhD	F/T	–	–	–	–
Nina	PhD	F/T	–	–	–	–
James	PhD	P/T	√	–	√	√

Pseudonym	Highest Qualification	Mode of Study	No First Degree	Part-time First Degree	Part-time Masters	Part-time PhD
Olivia	MSc	P/T	–	–	√	–
Polly	BSc	F/T	–	–	–	–
Tim	PhD	P/T	√	–	√	√
Rachel	PhD	P/T	–	–	√	√
Totals			12	6	24	14

Note: BSc = Bachelor of Science; MBA = Master in Business Administration; MSc = Master of Science; PhD = Doctor of Philosophy.

Index